BIRD
PERU

Clive Byers

BLOOMSBURY
LONDON • NEW DELHI • NEW YORK • SYDNEY

Bloomsbury Natural History
An imprint of Bloomsbury Publishing Plc

50 Bedford Square   1385 Broadway
London   New York
WC1B 3DP   NY 10018
UK   USA

www.bloomsbury.com

First published by New Holland UK Ltd, 2007 as
A Photographic Guide to the Birds of Peru
This edition first published by Bloomsbury, 2016

British Library Cataloguing-in-Publication Data
A catalogue record for this book is available from the British Library.

Library of Congress Cataloguing-in-Publication data has been applied for.

ISBN: PB: 978-1-4729-3216-7
ePDF: 978-1-4729-3217-4
ePub: 978-1-4729-3215-0

2 4 6 8 10 9 7 5 3

Designed and typeset in UK by Susan McIntyre
Printed in China

To find out more about our authors and books visit www.bloomsbury.com.
Here you will find extracts, author interviews, details of forthcoming events
and the option to sign up for our newsletters.

CONTENTS

Acknowledgements

I would like to thank all the photographers who have kindly supplied their collections of superb images for my selection. Many of their pictures are unique in terms of subject or quality. I have tried to include a cross-section of the work of James Walford, Andrew Lawson, Ron Hoff, Colin Bushell of Toucan Tours, Angus Wilson and Martin Hale. John Hornbuckle's images of rare species were particularly welcome. I am greatly appreciative to all these photographers. My thanks also go to Eustace Barnes and to John Brodie Good of Wildwings Travel Agency for putting me in contact with some of the photographers, to Jo Hemmings and James Parry at New Holland for commissioning me to write the book and for steering it through to completion, and last, but not least, to Barry Walker, owner of Manu Expeditions, who gave me the opportunity to spend a lot of time in Peru.

Acknowledgements for new images: 56 below NPL/Mary McDonald; 57 top, 130 top NPL/Visuals Unlimited; 63 bottom FLPA/Martin Hale; 66 top right NPL/Brent Stephenson; 68 bottom NPL/Pete Oxford; 69 below, 79 below FLPA/James Lowen; 84 below left, 91 below NPL/Tui de Roy; 88 below NPL/Nick Garbutt; 93 below Dario Sanches; 101 top FLPA/Neil Bowman; 117 below NPL/Tom Vezo; 126 below NPL/Luiz Claudio Marigo; 131 below NPL/Hermann Brehm.

Neotropical Bird Club

The addresses that are most helpful to anyone planning a birding trip to Peru are included in the introduction as websites or email contacts. However, those wishing to know more, or interested in becoming involved in data recording, should contact and join the Neotropical Bird Club (NBC). This is probably the most influential organisation involved with South American birds. Visit the club's website at www.neotropicalbirdclub.org or write to: The Neotropical Bird Club, c/o The Lodge, Sandy, Bedfordshire SG19 2DL, U.K.

INTRODUCTION

This book is designed to introduce you to the amazing birds of a remarkable country. Although it cannot tell you everything you need to know, it will certainly help you start to understand the complexities of birding in Peru and gain a good understanding of 252 of the most interesting species found there. Many of the photographs in the book are of common birds that you are likely to see in the environs of Lima or Cuzco or on a visit to one of the many archaeological sites, such as Machu Picchu. However, we have also included photos of a few rare, beautiful and elusive birds, which you are unlikely to see without professional help but which serve to underline the rich local bird diversity.

Peru is arguably the most biologically diverse nation on earth. Ecuador and Colombia to the north, Brazil to the east and Bolivia to the south also all lay claim to this title, but the sheer size of Peru and its position straddling the Andes just south of the equator mean that it is probably justified in its claim. Although Ecuador actually straddles the equator, it is so much smaller than Peru that it is unlikely to be as ecologically diverse overall, and whilst Brazil has an enormous land area compared with Peru, it has far fewer species, as it does not enjoy the presence and influence of the Andes or indeed that of the cold Humboldt current that introduces so many seabirds into the Peruvian picture.

Whilst forests in other parts of South America are subject to an almost uncontrolled process of deforestation, the lowland Peruvian forests are thankfully less seriously threatened. This is due largely to a homegrown conservation movement, which recognizes the importance of natural reserves and the potential economic value of ecotourism. These protected areas and the local employment possibilities they offer are increasingly acknowledged as part of a long-term answer to the future of the Peruvian people in a constantly changing social and economic environment.

To attempt to explain the fascinating and complex birdlife of this incredible country in a few paragraphs is an almost impossible task. The birds represented in this book are but a cross-section of a list of species that at the time of writing (September 2006) approaches a staggering total of 1870. Every year, new species are discovered and taxonomic changes mean that even more birds are added to the list as single species are spilt into two or even more. The species that are included in this book are largely representative of the many families of birds that occur in Peru.

A simplified explanation of the geography of the country helps provide an insight into the reasons why this country has so many bird species. The extraordinary biological diversity in this part of South America is partly due to the huge range of habitats occurring so close to the equator because of the presence of the Andean mountain range. This, combined with the variations in lowland forest habitats in the upper Amazon, the aridity of the Pacific slope, and the cold, food-rich Humboldt Current, add up to a multitude of habitat types supporting a fantastic variety of birds.

Most visitors arrive in Lima. This huge, sprawling city is situated on the coast at the northern end of the Atacama Desert, and the cold waters of the Humboldt Current in the south-eastern Pacific Ocean lap at its shoreline. Huge numbers of seabirds breed on offshore islands here, and were historically the focus of attention for those exploiting the massive amounts of guano produced by these birds. Further offshore there are many different species of migrant seabird, most of which breed on Pacific islands but pass through the region on their extraordinary journeys. The littoral zone teems with birdlife, with millions of shorebirds and gulls taking advantage of the feeding opportunities along Peru's beaches.

Inland from the shoreline the desert prevails. Some birds seem to thrive in this hostile environment, and can be found in places that are sometimes completely devoid of vegetation. In the north the desert gives way to a damper and more forested environment, where the cold current veers away from the coast and allows the warmer equatorial current to dominate the weather patterns. Increasingly regular El Niño manifestations can cause the desert to be flooded periodically. In the south, where some of the driest conditions on earth are the norm, these events are sometimes catastrophic.

River valleys and nearby irrigated areas provide birds with habitats of rich greenery in the form of orchards and fields of various crops. Inland of the coastal desert, the Andes rise sharply to dizzying heights, with the greatest peaks exceeding 6000m. Patches of Polylepis forest and scrub cling to the precipitous slopes and harbour a range of unique endemic birds.

The high plateau, known as the Altiplano, is also home to a number of bird species found only in Peru. 'Puna' grassland is the dominant habitat at these altitudes and is interspersed by fluvial run-offs – often referred to as 'bogs' – caused by glacial meltwater. Intermontane valleys also have their own microcosmic environments and associated avifauna.

On their eastern side the high Andes plunge into the Amazon basin, through wet cloud forest that descends for thousands of metres and provides a range of different birding possibilities at every elevation. This lush east slope could not be more different from the arid west slope. At the base of the Andes one emerges into the lowland Amazon rainforest, another world again. These forest habitats are amazingly diverse and harbour a huge variety of bird species. The two main forest types arc 'Varzea' forest, which is located on the floodplain and is seasonally flooded, and 'Terrafirme' forest, which occurs on higher ground above the flood levels.

HOW TO USE THIS BOOK

This book features 252 species of bird in detail. Each is illustrated with at least one colour photograph, showing the bird in its natural habitat in Peru. The associated text covers all the key information that you will need to make a positive identification. Useful pointers are given towards confusion species, and corner tabs will help you navigate the different bird families.

Photographs
Each species description is accompanied by at least one colour photograph. For many species, the plumages of the male and female are identical, and identification from the photograph should present

no problem. For others, however, males and females differ, and in such cases we have usually depicted the male but described the female in the text. Immature birds can present a problem to the beginner and expert alike. Some young birds (e.g. Harpy Eagle) can be quite different from adults, and most are drabber, with fewer diagnostic features. We have tried to cover these differences in the text.

Some species also have a different winter plumage from that worn in summer, the winter dress generally lacking the colour and finery of that when breeding. Here, we have aimed to use photographs of the plumage most likely to be seen in Peru.

Coloured tabs (see Key page 8)

The coloured tabs on each page show the major family groups of birds and provide the first clues to identification. The characteristic outline for each group will lead you to the section where you are most likely to find the species you have seen.

THE SPECIES DESCRIPTIONS

The descriptions give additional information to that provided by the photographs in order to help with identification.

Common name. In all cases, we have used the common or English name by which the species is most regularly known.

Scientific name. Each species has a Latin-based scientific name, recognized throughout the world.

Length. After the scientific name, the bird's approximate length, from bill tip to end of tail, is given in centimetres.

Range and status. Each species account includes an explanation of the bird's status in Peru (e.g. resident, summer visitor, passage migrant, winter visitor or vagrant), together with information on its preferred habitat on its level of abundance, and on how easy (or not!) it is to see.

IDENTIFYING BIRDS

Identifying birds can often be frustrating for the beginner, and sometimes even for the experienced birdwatcher. This can also be part of the fun of birdwatching: pitting your skills of observation against a creature that refuses to show itself clearly or, when it does, gives you just a back view as it flies away. Most of the birds in this book should be fairly easy to identify, provided a reasonable view is obtained. If you fail in this however, make a note of what you saw and you will be closer to identifying the species the next time you happen across it.

Your skills as a birdwatcher will improve as you gain experience. To aid this process, you should concentrate on the following aspects.

1. Size. It is not easy to gauge size so it is helpful to try to compare the bird in question with a known species. 'Larger than a sparrow', 'smaller

KEY TO COLOURED TABS

Tinamous

Pelicans to frigatebirds

Herons, egrets & storks

Ducks

Raptors

Gamebirds

Rails & allies

Waders

Gulls

Pigeons & doves

Parrots

Cuckoos & allies

Owls & nightjars

Swifts & hummingbirds

Trogons

Kingfishers

Motmots

Jacamars & barbets

Toucans & allies

Woodpeckers

Woodcreepers

Ovenbirds

Antbirds

Becards, tityras & cotingas

Manakins

American or tyrant flycatchers

Swallows

Pipits

Wrens

Thrushes & allies

Tanagers & honeycreepers

Blackbirds & orioles

Sparrows, finches & allies

8

than a crow' or simply 'a very large, tall bird' are all very useful starting points towards a correct identification.

2. Shape. Birds of particular families have a distinctive shape. For instance, herons and egrets are rather large, with long neck and legs, and all birds of prey have a hooked bill, while buzzards and eagles also have broad wings. In addition to size, it is especially important to note the shape of the bill and length of the legs. Some features, such as the presence of a crest or long tail streamers, will often enable an identification to be made immediately.

3. Colour. A knowledge of the feather group is the basis for describing a bird's plumage, and noting the colour pattern accurately is essential for making a correct identification. The stylized drawing below of a typical bunting shows these parts of a bird's plumage clearly, and if you want to progress in the skills of bird identification it is essential that you learn them. No matter how poor an artist you are, make a sketch of any bird you cannot identify and use this stylized drawing to note the colours and markings of the various parts of the bird.

4. Behaviour. Watch the way a bird feeds: does it search for insects, or is it a fruit-or seed-eater? Is it a tree-dweller or a ground-dweller? Does it fly fast or slowly? Is it found in flocks? These are all aspects that can assist greatly with identification.

5. Habitat. Every species of bird is adapted to live in a particular habitat, and this is an important clue to its identity. Most wader species, for example, are found on the shores of wetlands, including the sea, where they feed on invertebrates living in the mud. Nectar-feeders such as hummingbirds, however, require the presence of flowering plants and so are mostly confined to areas with abundant vegetation.

6. Voice. Some songs and calls are so distinctive that, once heard, they arc never forgotten. Most, however, can be quite subtle or can closely resemble those of several other species. While listening to tapes will help, there is no substitute for learning a bird's voice in the field.

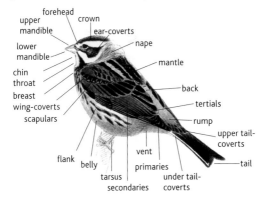

GOING BIRDWATCHING

While it is possible to watch birds without any equipment, binoculars will add greatly to your enjoyment. Later on, you may wish to purchase a telescope. These items of optical equipment can be very expensive and you should not rush into making a purchase until you are sure of what would best suit your needs. Before you make a decision, it is strongly advisable to test a range of binoculars and telescopes, ideally by noting what your local birdwatchers are using and asking to try them out.

There are however, a few simple guidelines. Never buy binoculars with a magnification of greater than x10, as it will be difficult to hold steady; ideally the magnification should be between x7 and x10. The size of the object lens should not be less than 30 or the light gathering will be poor. Most birdwatchers use 8 x 30, 8 x 40 or 10 x 40. Always consider the weight, as a heavy pair of binoculars can cause an aching neck and arms.

So far as telescopes are concerned the best advice is again to see what other birdwatchers are using and ask to look through them. Remember, too, that a sturdy tripod is important, as a flimsy one will move easily in the slightest breeze and this will annoy you intensely.

WHERE TO WATCH BIRDS IN PERU

When thinking of where to watch birds in Peru one may well be excused from wondering where to start. First and foremost, it is recommended that you invest in Thomas Valqui's book *Where to Watch Birds in Peru* (see Further Reading). This is an indispensable tome if you are planning to 'go it alone'. Thomas is perhaps Peru's most knowledgeable native ornithologist and his book will tell you where all the best birding areas are, as well as giving you a good idea of how exciting a birding trip to Peru will be.

Peru's incredible natural diversity means that the opportunities for birding are pretty much endless. One can start straight away by taking a taxi from Lima to the Pantanos de Villa, (the Villa marshes), just south of the sprawling city. Here you can find an array of birds on the wetlands, surrounding fields and along the shore. There is now a visitor centre and interpretive display, and guides are available. When birding close to urban areas it is advisable to be cautious and have at least one companion with you to dissuade anyone who may want to snatch your camera or binoculars. Taxi drivers are generally trustworthy as they are licensed and have identity cards and registration plates.

Also relatively close to Lima are great birding areas such as the Lamas de Lachay to the north and the Santa Eulalia valley for west-slope endemics such as the Great Inca Finch. This destination is not for the faint hearted, however. The precipitous slopes and the precarious single track 'road' that traverses them are something of an eye-opener for first-time visitors! Another option is to drive the main road inland which takes you to an altitude of over 4000m within three hours. Here one can find enigmatic birds such as Diademed Sandpiper-plover, White-bellied Cinclodes and Rufous-bellied Seedsnipe in the area of Milloc bog. However, one must be aware of the possibility of altitude sickness here.

1 Rafan (Mocupe)
2 Upper Maranon Valley (Balsas-Celendin-Leimabamba)
3 Abra Patricia
4 Chan Chan
5 Milloc Bog and Santa Eulalia Valley
6 Ballestas Islands and the Paracas Peninsula
7 Manu National Park
8 Upper Rio Madre de Dios and Amazonia Lodge
9 Pampas Del Heath
10 Colca Canyon
11 Laguna Salinas
12 Mejia Wetlands

Birding in rural areas is generally safe and the locals are, on the whole, welcoming, helpful and friendly. However, some areas in the north of the country, such as parts of the Huallaga valley, are still dangerous due to the activities of those involved in the cocaine business. The areas that are risky are well known but if in doubt, take advice.

Organized tours

In general, given the overwhelming number of birds and birdwatching sites in Peru it is advisable to engage the services of one of the many bird tour operators familiar with the country. This will save a lot of time and effort.

From the highland capital of Cuzco there are myriad birding possibilities. One can easily combine visits to archaeological sites, such as Machu Picchu, with great birding. The train to Machu Picchu drops steadily as it follows the course of the Rio Urubamba, along which it is worth looking out for Torrent Ducks and White capped Dippers. Cuzco is also the starting point for any expedition into the Manu National Park, one of the most extraordinarily diverse places on the planet but almost impossible to visit without the services of an agency. Although it is possible to rent a vehicle in Cuzco and drive yourself down the beautiful but rather frighteningly precipitous Manu road, it is recommended you take a bird guide and driver, as you will see a lot more birds this way. Travelling the road to Manu is an amazing experience for anyone, but especially for a birder. A further recommendation whilst in Cuzco is to contact Barry Walker at Manu Expeditions, which runs frequent tours to the Manu National Park. More information is available at www.manuexpeditions.com.pe or email manuexpeditions@terra.com.pe.

One of the best and most cost-effective ways to see a huge number of lowland forest species in a wonderful environment is to spend a few days at the rustic and family-run Amazonia Lodge, located on the upper Rio Madre de Dios at the lower end of the Manu road (www.amazonialodge.com). Nelly and Santiago Yabar and their sons are the perfect hosts and very keen to show their guests as many birds as possible. They have nearly 600 bird species on their list. On the way down the Manu road one can stay at the Cock-of-the-Rock Lodge, located in mjd-elevation cloud forest and where the fabulous bird that lends its name to the lodge is easily seen. The lodge can be booked via Manu Expeditions or in Cuzco.

Continue downriver to the Manu Wildlife Centre to sec yet more dramatic birds. This superb lodge affords the opportunity to see species which do not occur further upstream. Trips to this area are not cheap, however, the logistics for tour operators are complicated and expensive, and this is reflected in the prices charged for quality birding tours to the area. Equally recommended is Pantiacolla Lodge (www.pantiacolla.com), also located on the upper Rio Madre de Dios. One can also fly from Lima or Cuzco to Puerto Maldonado. This jungle town is the starting point for all trips to any of the many lodges along the Rio Madre de Dios and Rio Tambopata.

In the north there are a number of lodges quite close to the city of lquitos, which is situated on the upper Amazon. The varied forest types in this region include the 'white sand' forests with their own specific bird populations, very different from the forests in the south of the country.

Seasons and weather

As to be expected in such a diverse country, the weather in Peru is very variable and can be highly unpredictable. It is determined largely by geographical region, which can be broadly defined as the Andean highlands, the coastal plain, and the Amazon lowlands. There are effectively two seasons in Peru: wet and dry. In the highlands annual precipitation is quite low, with much of it falling between December and May, when there can be many wet days. Throughout the year, most mornings start sunny but cloud tends to build up during the day. Sunshine amounts average between 6–8 hours daily, and care must be taken to avoid sunburn. Days are generally warm, but the temperature falls quickly after sunset and frosts are commonplace during the dry season. The Peruvian coast has an unusual climate dictated by the presence of the cold Humboldt Current. Late December to March is generally hot and clear, but me rest of the year is dominated by fog and cloud, and the current keeps temperatures low, with little change between each month. Rainfall here is minimal. More unusually, the phenomenon known as El Niño can cause outbreaks of torrential rain, often with devastating consequences. To the east of the Andean range, the Amazon lowlands are characterized by consistently high temperatures and humidity for much of the year, and by heavy rainfall during the wet season (usually November to May). Conditions here are often uncomfortable.

FIELD TECHNIQUES FOR BIRDWATCHING

To the casual observer no equipment is necessary at all (apart from this guide book, of course!) but a pair of binoculars is invaluable in aiding identification. The next most important item of equipment is a notebook and pencil for taking field notes. When identification is in doubt, there is no better substitute for working systematically through a bird's characteristics and noting them down at the time of observation, or at least as soon as possible after the bird has flown off. These notes should include a description of the location (including habitat and altitude) and the bird's behaviour and call, as well as its appearance. Sketching the positions of markings such as stripes and barring can be a speedy way of noting characteristics in the field; it doesn't have to be a work of art for it to be useful when you have time to refer to a guidebook.

If you are keen to add to the number of species you are going to see, the early morning is the best time for bird activity. Rainfall followed by sunshine often stimulates activity, so observing after rain can be rewarding, too. It is important not to disturb wildlife while observing it, both for the sake of the wildlife and those that follow you. Always leave an area as you found it and obtain any necessary permission or permits before entering private or protected area. Keep to designated paths, move slowly and avoid making loud noises if you want to observe anything other than annoyed people. Remember to wear clothes of an appropriate colour to the environment – greens and browns are best.

GREAT TINAMOU *Tinamus major* 43cm

A large ground-dwelling bird, with cryptic plumage. Fairly common in various types of lowland forest east of the Andes, but more often heard than seen. Beautiful far-carrying calls accelerating to high-pitched trills are evidence of its presence. Most often seen crossing a path or forest roadway. Relatively large body and long, thin neck and head are typical of the bigger members of the tinamou family. The large eyes are an indication of its crepuscular life in the dark forests of the Amazon basin. Can be seen at Manu Wildlife centre on the Rio Madre de Dios. Occurs in both Terra firme and Varzea habitats.

HUMBOLDT PENGUIN *Spheniscus humboldti* 68cm

Taking advantage of the cold and food-rich Humboldt current, this aptly named bird is the second northernmost breeding penguin in the world after the Galapagos Penguin *Spheniscus mendiculus*. Breeds on offshore islands, notably the Ballestas Islands. Can sometimes be seen from the mainland shore at the Paracas peninsula, but is rather uncommonly observed in inshore waters and is only really guaranteed on or around the offshore islands. Occasionally seen in the water from boats, but luck is needed given the nature of the species; it is difficult to find them even in their breeding places, as they tend to come ashore late in the evening or early in the morning.

LEAST GREBE *Tachibaptus dominicus* 23cm

As its name indicates, this is by far the smallest of the seven species of grebe in Peru. It is nowhere common and is rather localized because of its dependence on particular wetland habitats. As with other grebe species, it builds a floating nest and the parents often carry their tiny striped chicks on their backs. The adults are delightfully diminutive and are characterized by their blackish plumage, particularly the head, which contrasts with the bright yellow eyes. Can be seen on oxbow lakes in the Amazonian lowlands and also in the extreme northwest in the Tumbes region.

GREAT GREBE *Podiceps major* 60cm

Fairly common along the coast. The largest and one of the most elegant and beautiful of the grebe family. Breeds commonly on the lakes of the Villa marshes south of Lima and other wetlands. As is typical amongst grebes, it builds a floating nest of vegetation, thus allowing for changing water levels. Many spend the winter months on the sea, when the beautiful slender neck, peaked head and long dagger-like bill are diagnostic. Its summer plumage is characterized by richer coloration, especially the long and deep reddish-coloured neck. This is very much a bird of the western coastal zone and does not occur in the fresh wetlands east of the Andes.

BLACK-BROWED ALBATROSS *Thalassarche melanophris*
83–93cm

The commonest albatross to occur in the waters of Peru after the Waved Albatross *Phoebastria irrorata*, which appears from the north. This is a much more elegant bird than that species and has a much smaller pale orange bill. With a wingspan of about 2.5m this is a very distinctive albatross, with blackish upperparts, white undersides and black-edged underwings. Its bright white head sets it apart from the rather similar Grey-headed Albatross *Thallasarche chrysostoma*, which also visits Peruvian waters but far less regularly. The Black-browed breeds on the islands off southern Argentina and the Falklands, but roams widely when not breeding.

WAVED ALBATROSS *Phoebastria irrorata* 83–93cm

This large albatross is frequently seen in a rather heavy and labouring flight, due to the fact that it only rarely has the advantage of the stormy conditions that most other albatrosses enjoy in the southern oceans. This gives it a strange appearance, and at a distance it can even resemble a Peruvian Pelican. It also has brown upperparts and whitish underparts, and the long neck and very large bill add to this general impression. Breeding on the Galapagos Islands and Isla de la Plata in Ecuador, they disperse south along the coast and can occasionally even be seen from land, particularly in stormy El Niño conditions.

SOUTHERN GIANT PETREL *Macronectes giganteus* 90cm

This Antarctic-breeding seabird, with a wingspan of over two metres, regularly visits Peruvian offshore waters in the non-breeding season, usually between April and September, most usually in the south. It is sometimes seen from boats visiting the Ballestas Islands. A large albatross-like bird, it comes in a variety of shades varying from almost white to almost black. A vicious predator, it has a massive bill which it uses to kill other smaller seabirds or scavenge almost anything edible that it can find. It often follows ships closely to take advantage of the slipstream, thus saving energy and enabling it to pick up anything edible brought to the surface by the ship propellers.

SOOTY SHEARWATER *Puffinus griseus* 40–50cm

An abundant migrant that passes Peru in offshore waters on its journey southward to its breeding areas on islands off southern Patagonia. Most likely to be confused with the bulkier White-chinned Petrel *Procellaria aequinoctialis*, but that species is a much plainer and uniform brown. Depending on wind conditions, the flight of the Sooty Shearwater is often very fast, its long, narrow wings giving it a distinctive profile. Although generally dark brown with a silvery flash on the underwing, it can look very different in various light conditions, sometimes appearing much paler in bright light. This species usually flies quite low over the water, but in windy conditions it can wheel high in the air.

PINK-FOOTED SHEARWATER *Puffinus creatopus* 50cm

This large shearwater is another long-distance migrant from the south that flies through Peruvian waters on its long journey up to the northern Pacific for the winter, and then again on its way south to breed. Most often seen on its southward journey, it is fairly common offshore whilst on this stage of its migration but rarely approaches the coast except in stormy conditions. Breeds on the islands off the southern tip of South America. It has a pattern which is quite distinctive among those seabirds occurring in Peruvian waters, being mid-brown on the upperparts and whitish below, when seen in good light. As its English name suggests, this species has pink feet, but these are rarely visible as it is nearly always seen in flight.

PERUVIAN DIVING-PETREL *Pelecanoides garnotii* 23cm

This little seabird is becoming increasingly rare, for reasons which are not yet fully understood. Its distinctive fluttering and rather inelegant flight, with fast flapping wings, is characteristic and resembles that of the Little Auk *Alle alle* of the northern hemisphere. It is also typified by its habit of diving into the water from flight, and also from the surface whilst swimming. When foraging underwater it uses its wings to 'fly' in a penguin-like fashion. This species is most often observed off the southern rocky coast, but it is still possible to see it from the Paracas peninsula. However, being very small, it is rather inconspicuous on the water.

HORNBY'S STORM-PETREL *Oceanodroma hornbyi* 21cm

This is one of the most beautiful of all the storm-petrels. It has very long and pointed wings with a pale stripe on the upper side. The most striking and obvious feature of this species are the white underparts and throat, separated by a distinctive charcoal-grey breast-band. These features contrast with the rather plain dark underwing. The breeding grounds of this enigmatic and rather mysterious storm petrel have still not been discovered. It is suspected that it will eventually be found to nest at a location somewhere in the arid mountains inland of the Atacama Desert. This species is only likely to be seen from boats far offshore, and it tends to be a rather uncommon bird in general.

MARKHAM'S STORM-PETREL *Oceanodroma melania* 23cm

This dark brown storm-petrel is a very typical member of this group of seabirds, being of a very similar pattern to a number of other members of the family. It has a light band across the upper side of the wings which is typical of the storm-petrels, but is otherwise a dark, plain brown bird. It is rather less long-winged than Hornby's Storm-Petrel *Oceanodroma hornbyi* and has a very different appearance from what is suspected to be a fairly close neighbour. Its nesting grounds were previously unknown, but it has recently been found to breed in the Atacama Desert of the south. This species is quite commonly seen off the coast of Peru, usually quite far offshore.

MASATIERRA PETREL *Pterodroma defilippiana* 27cm

Also known as Defillipe's Petrel, this is an uncommon but quite regular visitor to the waters of the Humboldt current off the southern coasts of Peru, breeding on the Juan Fernandez Islands of Chile. It is characterized by its mostly grey plumage, white underparts and long wings with a black band across the upper wing. As is usual with the Pterodroma group of petrels, in windy conditions it typically flies with a high wheeling action. However, it is only one of a group of seabirds with a similar plumage pattern that might occur in Peruvian waters. The black mask, however, is most obvious in this species. Could be confused with the Juan Fernandez Petrel *Pterodroma externa*, but that bird is larger, even longer-winged and has a less obvious black mask.

WHITE-CHINNED PETREL *Procellaria aequinoctialis* 56cm

This large, dark brown petrel is a fairly common visitor from the far south, where it breeds on the sub-Antarctic islands. The white chin is very difficult to discern at the best of times and may be barely present in some birds, but the whitish bill is always a prominent feature of this species. Although very similar to several other *Procellaria* species, this is by far the most common in Peruvian waters. Can be separated from the Sooty Shearwater *Puffinus griseus* by its heavier profile and lack of pale flashes on the underwing, and by appearing generally dark chocolate-brown both above and below. Also tends to fly with a less energetic action than that species.

CAPE PETREL *Daption capense* 40cm

Also known as the Pintado Petrel or 'Painted' Petrel, due to its attractive plumage pattern. This is basically a black-and-white bird, speckled on the mantle and wings, and with a black head and white underparts. Another of the Antarctic-breeding seabirds which utilize the nutrient-rich waters of the Humboldt current during the Austral winter months, it occurs only uncommonly as far north as Peru. Often follows ships closely and is fondly regarded by seafarers, who traditionally called it the 'Cape Pigeon', its plumage pattern reminding them of the Feral Street Pigeons *Columba livia* of their home cities.

RED-BILLED TROPICBIRD *Phaetheon aethereus* 60cm

This elegant seabird is an uncommon visitor. It breeds both to the north and to the south of Peru but not in Peruvian waters and, although rarely encountered, is probably more regular than is presently known. Mostly likely to be seen in the south, as I have personally seen this species far offshore there, probably a bird from the Chilean Juan Fernandez islands, where it breeds. Its mainly white body, black on the wings and long white tail make it very distinctive and easily recognizable. Although this is normally a tropical seabird, birds naturally take advantage of the rich pickings of the Humboldt current, occasionally appearing off Peru.

BLUE-FOOTED BOOBY *Sula nebouxii* 86cm

Most often encountered off the northern shores of Peru, where it is quite common, this member of the family is a migrant from its breeding grounds off Ecuador. The main feature of the species are of course its remarkable blue feet, although they are rather difficult to see when the birds are in flight. Can be confused with the Peruvian Booby *Sula variegate*, but that species has a pure white head whereas the Blue-footed Booby has a browner head. It can however be confused with immatures of that species, which lack this normally shining white feature.

PERUVIAN BOOBY *Sula variegata* 73cm

The commonest member of the booby family in Peru and one of the guano-producing birds that have historically been so important to the economy of Peru. It is easily identifiable by its white head, scaly pattern on the dark brown upper side and its white underparts. Unlike the Nazca Booby *Sula granti*, it usually fishes in large flocks, being a much more common bird in Peru, normally far offshore, diving at great speed to carry it several metres underwater. Breeds in large colonies on the islands off the coast, but no longer in the vast numbers that used to feed on the multitudes of anchovy that were formerly available.

NAZCA BOOBY *Sula granti* 73cm

Only recently reclassified as a full species distinct from the Masked Booby *Sula dactylatra*, which is very similar but has a yellow bill. This species has a pinkish-orange bill. It is another of the seabirds that visit the coast of Peru from breeding areas in the Galapagos Islands. Named after the Nazca area of the southern coast of Peru, this booby is an expert fisherman, often accompanying ships to ride the slipstream and catching flying fish in the air as they are flushed from the sea by the vessel. Easily recognized by its white body plumage and the contrasting black on the wings, which is quite unlike both the Peruvian Booby *Sula variegata* and Blue-footed Booby *Sula nebouxii*, which have brown upperparts.

PERUVIAN PELICAN *Pelecanus thagus* 152cm

One of the most obvious and distinctive birds of the Peruvian littoral. Also the largest of the guano-producing species, it breeds in large numbers on offshore islands. Often seen taking advantage of fishing boat harbours, such as Salaverry and Lima, where they will crowd onto the roofs of the fish markets to watch for waste fish being discarded and scavenge anything thrown out – a very opportunistic bird in this respect. Frequently seen flying to and from their breeding colonies in flocks in 'V' formations typical of very large birds. Once in the air they are graceful flyers, but after gorging at the fish market they sometimes have trouble getting into the air at all.

NEOTROPIC CORMORANT *Phalacrocorax brasilianus* 70cm

This is one of the commoner aquatic birds along the rivers of the interior but also occurs along the coast. Also called the Olivaceous Cormorant, the adults indeed have a deep greenish-brown hue, although they can appear generally blackish in normal conditions. Can be confused in its inland range with the Anhinga *Anhinga anhinga*, but that bird has a much longer neck and bill, pale upper wings and swims with most of its body submerged, whereas the cormorant usually swims with its body higher in the water. This species regularly gathers to feed in particularly good fishing points along the lowland rivers. All the other cormorants of Peru are strictly coastal.

GUANAY CORMORANT *Phalacrocorax bougainvillii* 76cm

Once the commonest of the guano-producing seabirds of Peru, its populations suffered a catastrophic collapse due to the over-fishing of anchovies, the mainstay of its diet. This was one of the mistakes that led to the demise of both the guano and fishing industries. The population of this bird dropped from tens of millions to just a few million, but the species has made a significant recovery in recent years. Around the Ballestas Islands, where it still nests in large colonies, it can be seen fishing and flying in huge flocks containing many thousands of individuals. Black above and white below, it is clearly recognizable.

RED-LEGGED CORMORANT *Phalacrocorax gaimardi* 70cm

This cormorant is completely different from the other cormorant species, having a very distinctive grey plumage with white-spotted wings. It also has large white patches on the sides of the neck, a very elongate yellowish bill and, as its name indicates, bright red legs. The appearance of this cormorant is unique among the seabird community of Peru. It is not a colonial breeder like the Guanay Cormorant *Phalacrocorax bougainvillii*, but nests in different places along the cliffs of islands such the Ballestas, where its very unusual plumage makes it rather conspicuous. The young birds are less obvious, being rather duller and lacking the spectacular plumage of the adults.

MAGNIFICENT FRIGATEBIRD *Fregata magnificens* 106cm

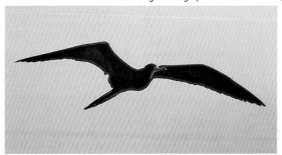

This large and very distinctive bird is common along the northern Peruvian coast. Breeds commonly in Ecuador. It has a very distinctive profile with a long hooked bill, long pointed wings with a span of over 2.5m, and a long forked tail. The female has a white neck and breast but the male has a bright red inflatable sac on his throat. A predatory species, it regularly attacks other seabirds to make them regurgitate their last meal. Also picks food from the water's surface by swooping down, but does not land on the water. Unless breeding it spends nearly its entire time in the air, except at night when it roosts in trees. At times it will soar to amazing heights, at which it is barely visible to the naked eye.

COCOI HERON *Ardea cocoi* 127cm

Also known as the White-necked Heron, this is the commonest large heron along the rivers of the eastern lowlands. Very tall and conspicuous, especially when its long white neck is extended. The adults have a neat black cap. Upperparts are blue-grey and underparts a combination of black and white, the black belly contrasting sharply with the white breast. Young birds are more uniform grey, lacking the distinctive white neck of the adults. Appears very large in flight, having expansive wings and a rather elegant, slow, flapping flight action. It occasionally appears on the coast and also in the highlands, but is generally a distinctive bird of the Amazonian region of Peru.

GREAT EGRET *Ardea alba* 96cm

This is the large white heron which occurs commonly in wetland habitats along the coast and in the eastern lowlands. Also quite regularly seen in many agricultural areas and other places such as the wetter areas in urban zones, where it is typically opportunistic. Sometimes gathers in large numbers to take advantage of particular feeding opportunities, often with other members of the family. As is also typical of the herons, it nests colonially in trees. Conspicuous by its large size, its pure white plumage and very long neck and bill. Has a lazy, rather elegant flight action.

SNOWY EGRET *Egretta thula* 58cm

This common but much smaller relative of the Great Egret *Ardea alba* is another of the twenty species of the heron family in Peru. Like its larger congener, it occurs in both North and South America, and is also pure white but, as it has a much smaller bill, it probably tends to hunt for smaller prey items. However, one must be careful of confusion with the immature plumage of the Little Blue Heron *Egretta caerulea*, which is also pure white but has uniformly greenish legs. This egret has distinctive blackish legs and bright yellow feet. It displays long white plumes on the head when in breeding plumage, as well as typical *Egretta* plumes from its back.

TRICOLOURED HERON *Egretta tricolor* 60cm

A locally common heron in coastal marshes, this handsome and rather graceful species can be seen in various wet-land localities, including marshes such as Las Lagunas de Mejia and at Pisco, also south of Lima and close to Paracas. It is characteristic in having a white belly and throat, a deep grey neck and purple-rufous tinged back. These are the three main colours that lend this lovely heron its English name. It can be a stealthy hunter and at other times quite energetic in chasing and capturing prey. It occurs widely through both North and South America.

LITTLE BLUE HERON *Egretta caerulea* 58cm

Adult

Immature

The 'blue egret'. This bird is quite a common sight in both fresh and saline wetland habitats all the way along the coast. The adult plumage is a lovely mix of maroon on the head and neck with deep blue-grey coloration to the mantle and wings. The immature, however, is completely different and somewhat peculiar, in that it is almost totally white and closely resembles the Snowy Egret *Egretta thula*, except in having plain greenish legs and lacking yellow feet. Often hunts in a very energetic fashion by actively pursuing its prey but, as with most herons, it will also adopt a 'stand and wait' strategy.

BLACK-CROWNED NIGHT-HERON *Nycticorax nycticorax* 63cm

Adult

Immature

This is quite a common heron in the Puna zone of the Andes and also occurs along the coast, although it much less common there. Interestingly, in most parts of the world this shy and retiring species is largely crepuscular or nocturnal and so rarely seen during the day. However, in the highlands of Peru it can often be seen foraging in the open, sometimes at the side of busy roads, even at very high altitudes. The adult has a striking appearance with its black upperparts and lovely dove-grey body plumage. White plumes hang from the back of its head. The immature is very different in appearance, being a dull brown with streaks and spots.

YELLOW-CROWNED NIGHT-HERON *Nyctanassa violacea* 70cm

This species is a relative of the Black-crowned Night-Heron *Nycticorax nycticorax*, but unlike its cousin it does not occur in the highlands or feed in daylight by the side of busy roads. It is restricted to the north coast in Peru, in particular to the mangroves along the Tumbes coast. The adult has a striking head pattern which is unlike that of any other bird, having black-and-white stripes and, as its name indicates, a yellow crown. Its generally grey body plumage is also rather attractive. The immature bird has a very different plumage, composed of streaks and spots, and rather similar to that of the immature Black-crowned Night-Heron *Nycticorax nycticorax*, although a richer and more rufous colour than that species.

STRIATED HERON *Butorides striatus* 40cm

This small heron is rather bittern-like and a very close relative of the Little Green Heron *Butorides virescens* of North America. It is a local and rather uncommon bird in Peru, but occurs on both sides of the Andes. Found in wetlands along the coast and anywhere there is suitable habitat in the eastern humid lowlands and foothills, occasionally up to about 1000m elevation. The bird's English name is confusing, as it does not look particularly striated and is in fact a rather plain dark brown and grey-green; the 'Striated' element is presumably derived from the pale edges of the wing feathers. It appears to undertake some altitudinal movements and has been seen quite regularly on the altiplano at an altitude of 4000m.

LEAST BITTERN *Ixobrychus exilis* 26cm

The tiniest of the herons, the Least Bittern is a very agile and stealthy bird, sometimes adopting a hunting posture which sees the bird hang almost upside down to hunt from above. It has white stripes on its back and less obvious striations below. The male is mostly black above and the female brown and chestnut. Although a very handsome bird, it is often overlooked due to its small size and cryptic plumage. It is quite common on the coastal plain, and at home anywhere there are marshy habitats with reeds. The species also occurs in the eastern lowlands. The photograph shows a female or immature individual.

FASCIATED TIGER-HERON *Tigrisoma fasciatum* 66cm

One of the more unusual herons, in that it is found in quite specific habitat: along fast-flowing rivers and streams along the eastern base of the Andes, and also less commonly in the intermontane valleys. Forages in the open, usually on gravel or rocky islands without vegetation, where its plumage allows it to remain remarkably inconspicuous. It adopts the 'stand and wait' strategy, looking out for something to come its way. Perhaps most easily seen on the upper Rio Madre de Dios in Cuzco province, where it can often be approached closely by boat. Its English name relates to the fine vermiculations or 'fasciations' on the neck. The young are boldly marked with black and buff 'tiger stripes' and very similar to young Rufescent Tiger-Herons *Tigrisoma lineatum*.

RUFESCENT TIGER-HERON *Tigrisoma lineatum* 70cm

Although closely related to the Fasciated Tiger-Heron *Tigrisoma fasciatum*, this species is much more likely to be seen on oxbow lakes and along smaller streams in forested habitats. Often only seen when flushed, it will at times perch in prominent places to catch the early morning or late evening sun. Like the Fasciated Tiger Heron *Tigrisoma fasciatum*, the plumage of the immature bird, with its bold buff and black chequered pattern, is very different from the adult and could easily be mistaken for a completely different species. The adult is a beautiful and finely vermiculated bird, with a rufous neck that is usually held retracted but can be extended to an astonishing length.

JABIRU *Jabiru mycteria* 132cm

Mostly white with a very large bill. Its black and pink bare-skinned neck is an obvious feature of this bird and is inflated in display, its name deriving from an American Indian word for 'to blow up with the wind'. One of the largest flying birds in Peru, it is easily recognized by its large size and not the best bird to encounter in a small plane whilst flying over the Peruvian rainforest! Fortunately it has very large white wings, about two and a half metres in span, which make it easy to spot. Rather uncommon and local, but seen more often in particular areas such as Pampas del Heath in the south-east or along the Rio Madre de Dios in Cuzco province, where it can be seen foraging or scavenging along the banks, and sometimes competing with vultures for carrion.

WOOD STORK *Mycteria americana* 100cm

Adult *Immature*

The smaller and commoner relative of the Jabiru *Jabiru mycteria*. With a wing span of over two metres, it is, like the Jabiru, a danger to small aircraft, as it sometimes soars in flocks to heights of 1000m or more. Most easily distinguished from the Jabiru by having black wing tips. Its black head and neck are distinctive but, unlike the larger Jabiru, the neck does not inflate. The bill is a completely different shape from that of its larger cousin, being smaller, slimmer and down-curved. Most often seen along the riversides, marshlands and pampas of the eastern lowlands, but is generally a rather uncommon bird in Peru, as it prefers more open, lowland habitats.

CHILEAN FLAMINGO *Phoenicopterus chilensis* 120cm

A commonly seen bird along the southern coast in suitable shallow water habitats, especially at Paracas. Also occurs quite commonly in the highlands on saline and freshwater lakes alike, particularly in the south of the country. Its greyish legs and contrasting red knees distinguish it from the other two Peruvian flamingo species, Andean *P. andinus* and James *P. jamesi*, which have different leg colours. Those species also have different bill patterns. Occurs on the Laguna Salinas above Arequipa alongside both other species, giving an opportunity to compare them all in one place. Feeds in typical, very specialized flamingo fashion, by sieving micro-organisms from the water with its highly adapted bill. Often migrates at night to avoid predation by diurnal raptors.

ANDEAN IBIS *Theristicus branickii* 58cm

As its name suggests, this ibis is an inhabitant of the highlands. As with the Andean Lapwing *Vanellus resplendens*, it is often found in the vicinity of human habitations, attracted by local irrigation practices. Usually found in any marshy habitats at high elevations. It is a large and thick-set ibis, and a noisy one too. It is generally grey above and buff below and is the only bird with this appearance at these altitudes. Although it closely resembles the Black-faced Ibis *Theristicus melanopis*, this is a rare bird of the Peruvian lowlands. Can be seen from the road from Cuzco to Paucartambo and surrounding areas, although it is commonest at higher elevations up to about 5000m. It shares its habitat with the Puna Ibis *Plegadis ridgwayi*, which is smaller and looks completely different, being mostly blackish.

ROSEATE SPOONBILL *Ajaia ajaja* 80cm

The only spoonbill species in the Americas, this large and rather strange pink bird with its preposterous bill and bare head is usually seen along the rivers of the eastern lowlands, but has been recorded on the coast in Tumbes and even at the Villa marshes south of Lima. It is totally unmistakable as there is no other bird that resembles it. The three flamingo species are the only other pink or pinkish birds in Peru but are a completely different shape. As with other members of the family, it uses its extraordinary spoon-shaped bill to sieve small invertebrates from the water. Quite commonly seen along the lower Rio Madre de Dios and on lowland lagoons.

HORNED SCREAMER *Anhima cornuta* 90cm

One of three species of a strange family, looking like a cross between a goose and a raptor. They will forage in wet habitats and perch conspicuously on tops of trees. It is a very large inhabitant of the eastern lowland rainforests, and has a loud call – hence its name. The 'horn' is actually a single plume, which protrudes from the front of the bird's head. Sometimes soars elegantly over the forest in a rather unlikely fashion for such an ungainly looking bird. More usually seen flapping hard to get its sheer bulk airborne. Forages in wet, forested areas in the eastern lowlands. Often seen on oxbow lakes or riverbanks, such as the upper Rio Madre de Dios.

ANDEAN GOOSE *Chloephaga melanoptera* 76cm

A bird of the puna zone, this black-and-white goose is quite unmistakable and a unique bird in this habitat. Its graphic plumage is diagnostic as there is no other similar bird in its highland range. It occurs in summer all the way up to the snowline, reaching 5000m, but migrates altitudinally in the winter, dropping to lower elevations to escape the snow and ice. It is a common bird at these high altitudes and easily seen, as it is so clearly recognizable. Grazes on the abundant

pasture at these elevations and often competes with the also abundant llamas for access to this resource. Being a goose, it is often successful in fights and in forcing the animals out of their territories!

TORRENT DUCK *Merganetta armata* 38cm

One of the most easily recognized ducks. Only found along fast-flowing montane and submontane streams and rivers, it requires unpolluted water but is quite common where these conditions exist. Very easily seen along the Rio Urubamba, especially from the train to Machu Picchu between Ollantaytambo and Aguas Caliente, where it is possible to see more than ten pairs along the route. The male has a black-and-white striped head; this is very different from that of the female, which has a much plainer rufous and grey pattern. The precocious ducklings are able to swim at an early age and are sometimes seen negotiating rapids with great skill.

CINNAMON TEAL *Anas cyanoptera* 40cm

This is a common duck along the coast wherever there are suitable wetland habitats, and also in the highlands. It is commonly seen at Huacarpay lakes near Cuzco and even occurs at higher elevations on isolated ponds up to well over 4000m. The male has a beautiful, rich cinnamon body plumage, giving the bird its English name. The female is, as is usual in ducks and many other birds, much less colourful. Perhaps the most obvious feature is the light blue forewing and green speculum, which are apparent as soon as the bird takes flight. This feature is present in both sexes. It is a bird that occurs all the way from Canada to the Falklands.

SPECKLED TEAL *Anas flavirostris* 40cm

A lovely if not very colourful little duck with a green wing speculum, also distinguished by its bold pattern of dark head and pale underparts, speckled breast and bright yellow bill, which it shares with the Yellow-billed Pintail *Anas georgica*. That species, however, has a much more uniform appearance and is rather larger, longer and slimmer. Speckled Teal is quite common on highland lakes as high as 4500m, and common at Huacarpay lakes near Cuzco, but can also be seen on the coast in the winter months at certain wetland locations, such as the marshes at Villa, just south of the capital.

CRESTED DUCK *Anas specularioides* 60cm

A large and distinctive duck with a dark bill, commonly seen on lakes and ponds in the highlands, very often above 4000m elevation. It has large smudgy spots on its flanks and a white trailing edge to the wing. Also has a beautiful purple-pink wing speculum, this being the bright iridescent patch on the inner wing, which many ducks of the genus *Anas* have but which is often not obvious. It also has a distinctive head shape, although the 'crest' is not always immediately apparent. The dark patch on the side of the head highlights the bird's eyes. Occurs all the way south to Tierra del Fuego, where it can be found at sea level.

ANDEAN CONDOR *Vultur gryphus* 106cm

One of the largest flying birds in the world, with a wingspan of three metres or more, this is one of the most enigmatic birds of the high Andes. In rural areas it is often captured for use in traditional ceremonies. It is not a particularly common bird in Peru but can be seen from the train between Cuzco and Juliaca, particularly in the region of Abra La Raya. Best seen at the viewpoint overlooking the Colca canyon in the south of the country. Many supposed sightings of this species are misidentifications of the much commoner Black-chested Buzzard-Eagle *Geranoaetus melanoleucus*. However, the condor will never display a white belly in any plumage. The young birds do not have the white collar that so obviously distinguishes the adults.

KING VULTURE *Sarcoramphus papa* 76cm

Often referred to as 'El Condor' by lowland indigenous people, this is the largest and most powerful of the lowland neotropical vultures. It utilizes the abilities of other vulture species, particularly the yellow-headed vultures, to literally sniff out carrion. By watching them home in on the source of the smell it can then follow them down, muscling in to take the greater share of the spoils. It can then be seen enjoying itself as the other vulture species stand around waiting for it to leave, if indeed, by the time it has finished its meal, the bird can even get off the ground. Its large size, multicoloured head and mostly white body plumage set it apart from all other New World vultures.

BLACK VULTURE *Coragyps atratus* 63cm

The commonest sedentary member of the vulture family. It is the vulture most likely to be seen scavenging in urban areas, and is often allowed to do so undisturbed, as in some places there may be no other refuse collection service! It is very different in general appearance from other vultures and so is easily recognized, being the only vulture with a wrinkly black head. It also has a somewhat ungainly flight on rather short wings, as opposed to the lovely tilting and swooping flight of the Turkey Vulture *Cathartes aura*. It is quite voracious and on the ground it can run fast and successfully fight for the spoils with other vulture species, often simply overwhelming them by sheer numbers. A flock of Black Vultures in action can be a rather unpleasant spectacle.

GREATER YELLOW-HEADED VULTURE *Cathartes melambrotus*
75–80cm

This vulture is very typically a bird of the lowland eastern rainforests, where it is the commonest vulture soaring above the trees. It rather closely resembles the Turkey Vulture *Cathartes aura*, but is somewhat broader winged and has dark inner primaries, which that species lacks. Perhaps the most interesting feature of this bird is its ability to locate carrion in dense forest. It does this with its extraordinary sense of smell, which is said to be perhaps the most acute of any bird. Other vultures will watch this bird and follow it, taking advantage of its greater olfactory capability, and will then home in on the food source and compete with this species. It closely resembles the Lesser Yellow-headed Vulture *Cathartes burrovianus*, but that smaller relative scavenges over open savannah.

TURKEY VULTURE *Cathartes aura* 66–75cm

One of the most ubiquitous large birds to be seen in the skies over Peru, and often mistaken for a condor or an eagle. A scavenger, it is identifiable by its quite distinctive profile in flight, black plumage and, at close range, its variably coloured head. There are two quite different populations in Peru: one is native and the other consists of migrants from North America. It has a rather elegant, lilting flight on long wings, as opposed to the flapping, short-winged Black Vulture *Coragyps atratus*. It is very similar in general appearance to the two yellow-headed vulture species of the savannahs and rainforests, but is always recognizable by its multicoloured pink, yellow and purple head. Unlike those species, it will very often be seen over urban areas.

WHITE HAWK *Leucopternis albicollis* 55cm

This beautiful and aptly named raptor is a denizen of the lower slopes of the Andes. Rarely seen except when perched up, sunning itself in the early mornings, it is more often encountered soaring over its forest habitat later in the day. Like most other raptors, it will take to the wing when there are sunny conditions allowing thermals to form. This is often about mid-morning on a clear day. The species is mmediately recognizable by its mostly white body plumage, pure white head and very broad white wings, with a complex pattern of black toward the tips. Although conspicuous, this can be a hard bird to find and chance encounters are the rule.

BLACK-COLLARED HAWK *Busarellus nigricollis* 48cm

A medium-sized hawk of the eastern forested lowlands, where it occurs almost exclusively along rivers and lakes, feeding mostly on aquatic creatures such as crabs. Fairly common in these habitats, where it is usually seen perched on waterside trees. Very distinctive, with its reddish coloration and pale, almost whitish head and black breast-band. Often seen from boats travelling along the lowland waterways, as it soars over its riverine habitat. Very easily recognized by its distinctive plumage pattern, and because it often sits quite low and rather obviously on dead snags by the riverside.

SAVANNAH HAWK *Buteogallus meridionalis* 60cm

This is quite a large, broad-winged and short-tailed raptor, a specialist of the open country, as its English name suggests, occurring in grassland or marginal habitats such as cleared areas and riverbanks, and also in agricultural areas. Spends much time perched in the open, often on the ground. This hawk's vermiculated brown, grey, buff and rufous plumage appears quite subtle and under-stated when seen from a distance, but is quite beautiful at close range. Has a white-banded tail and in flight also shows cinnamon patches on the wings. Occurs on both sides of the Andes, but usually in the north on the west side and in the south on the east side.

ROADSIDE HAWK *Buteo magnirostris* 38cm

One of the most commonly encountered raptors on the eastern Andean slope and in the lowlands, where it is frequently seen in cleared areas and particularly at roadsides, which of course gives this hawk its very apt English name. In some places it would be better known as the riverside hawk, as it is regularly seen from boats when perched up in riverside trees. It has a very distinctive, rather inelegant 'flappy' flight action, due to having quite short wings. These display an obvious rufous patch in the primaries. Overall, adult birds are grey-brown above and barred on the breast and belly, whilst the immatures are browner, as illustrated in the photograph. The species has a distinctive mewing call, which is commonly heard where this bird occurs.

CRANE HAWK *Geranospiza caerulescens* 53cm

Occurs on the west slope in the north and rather less commonly in the lowland forests of the Amazon basin. A medium-sized and rather slim, slate-grey raptor with a banded tail and long red legs, which it uses to raid the nests of other birds and predate their young. This hawk is perhaps best recognized and differentiated from several other quite similar species by the whitish crescentic patches on the under primaries, which are apparent in flight. It is also a slighter bird than the other Peruvian grey hawks, such as the Plumbeous Hawk *Leucopternis plumbea* from Tumbes and the Slate-coloured Hawk *Leucopternis schistacea*, which also shares its habitat in the eastern lowlands.

BLACK-CHESTED BUZZARD-EAGLE *Geranoaetus melanoleucus* 68cm

Living in the high Andes, and being rather common, this large and distinctive raptor is often pointed out to visitors in places like Macchu Pichu as an Andean Condor *Vultur gryphus*. However, although it is a very large and obvious bird of prey in the air, it is actually an unmistakable bird in its own right, given its extra-ordinary shape and pattern. The combination of very broad wings and a very short tail should make its identification quite straightforward. The white belly and contrasting black breast of this species are also diagnostic. The Andean Condor has a longer tail, and longer and narrower wings with prominent 'fingered' flight feathers, and so presents a very different flight profile.

SHORT-TAILED HAWK *Buteo brachyurus* 43cm

This rather squat, broad-winged and short-tailed hawk is aptly named and commonly seen soaring over the lower foothills of the eastern Andean slopes up to an elevation of about 900m. It is a bird of the forests of these low elevations, also occurring in the forests of Tumbes in the north of the country. The species has two distinct forms, one with white underparts and white tail with a dark terminal band,

the other form being mainly dark brown with a lighter tail, again with a dark band at the tip. It is one of the most easily recognized of Peruvian raptors by its distinctive shape in flight. Occurs all the way from Central America to Argentina.

VARIABLE (RED-BACKED) HAWK *Buteo polyosoma* 45–55cm

Generally not a forest bird, this raptor is found in a wide range of montane habitats and is a common raptor of the puna grasslands, where it can be seen hovering in search of prey. It will sometimes come down to lower elevations, particularly on the western side of the Andes. The species varies greatly in terms of size, plumage type and coloration, and there is considerable confusion as to whether this bird is actually more than one species. However, most adults have a reddish mantle, which gives this hawk its English name, and adults also always have a white or pale tail with an obvious black band at the end. The form illustrated is the commonest type.

ZONE-TAILED HAWK *Buteo albonotatus* 50cm

This long-winged and elegant black hawk is found both in the forests of the eastern lowlands and the foothills of the Andes. It is a species that causes some confusion, as it bears a rather unlikely and unusual resemblance to the Turkey Vulture *Cathartes aura*. Its dark coloration and habit of cruising on uplifted wings with frequent tilting, much like the vulture, is the cause of this confusion. However, with a good view one should be able to see clearly the banded (zoned) tail, the finely barred flight feathers and the yellow area of bare skin at the base of the bill, which is known in raptors as the cere. Athough a rather uncommon bird, it is almost certainly overlooked and under-recorded because it is mistaken so often for the Turkey Vulture.

HARPY EAGLE *Harpia harpyja* 94cm

Adult (left); immature (above)

Being both an iconic bird of prey and the largest and most powerful raptor in Peru, this is unsurprisingly one of the most sought-after species by birders. Huge bill and talons enable this eagle to take large prey such as monkeys and sloths. The adult looks rather similar to the Crested Eagle *Morphnus guianensis*, but that species is smaller and less powerful. Harpy Eagles require areas of lowland forest that are able to support the quite large animals that they prey on. They are highly revered by indigenous people, but unfortunately are sometimes captured and kept as symbols of power. A pair will rear only a single chick once over a two-year period. The young birds have a white head, and so look very different from the adults.

SWALLOW-TAILED KITE *Elanoides forficatus* 60cm

This beautiful and elegant raptor with long pointed wings is one of the more conspicuous and most beautiful birds of prey in Peru. It is instantly recognizable, with its long forked tail, grey and black upperside, gleaming white head and underparts, and its striking black-and-white underwing pattern. Many thousands migrate from northern climes to spend the winter on the lower eastern slopes of the Andes, where it is often seen. However, there is also a sedentary resident population. Nearly always seen soaring over the forest and swooping down, picking small snakes and other prey items from the canopy foliage whilst in flight. Migrants have been recorded crossing the Andes at nearly 5000m elevation.

PEARL KITE *Gampsonyx swainsonii* 23cm

One of the smallest raptors of the region, even smaller than the American Kestrel *Falco sparverius*. It is grey above and white below, with a dark cap and white and orange collar. This very beautiful little kite is an uncommon bird of the northern coastal regions, and also the drier and largely more open regions on the east side of the Andes. Often perches on manmade vantage points, such as pylons and power lines, in the deforested habitats it prefers. Occurs up to an elevation of about 900m and feeds rather predictably on very small prey items. Given the amount of deforested habitat in some areas of Peru, it is surprising this species is not more common.

SNAIL KITE *Rostrhamus sociabilis* 36cm

Adult   Immature

This highly specialized raptor has a bill specifically designed to deal with the freshwater snails that are its sole food source. Lives along the rivers and oxbow lakes of the eastern lowland rainforests. Typical of the kite family, it is a beautiful and elegant bird. However, the uniform dark grey adult is very different from the immature bird, which has brown upperparts and is whitish and streaked below and on the head. The species can be quite common where suitable wetland habitat exists, and can very often be seen flying in typical 'floating' kite fashion over the lagoons in which it finds its very particular prey.

LINED FOREST-FALCON *Micrastur gilvicollis* 32–38cm

A denizen of the lowland Amazon forest, rarely seen but quite often heard calling at dawn and dusk. All the forest falcons have a distinctive yelping call but are rarely heard calling during the day. Vocalizing is very typical behaviour for this family of rainforest-dwelling raptors, and is often the only indication that forest-falcons are present at all. As with the other four species of forest-falcon in Peru, and as its English name suggests, this species is an adept forest hunter with the agility to skilfully negotiate the dark understorey at high speed. Hunts many species of small forest bird and, as its large white eyes attest, can do so in the low light conditions of the sometimes almost closed canopy.

BLACK CARACARA *Daptrius ater* 48cm

A fairly common and widespread raptor of the Amazonian rainforests east of the Andes, its loud and coarse call often alerts one to its presence. However, it does not make the protracted chachalaca-like calls uttered by its close relative, the Red-throated Caracara *Daptrius americanus*, which shares its habitat. That species also differs in having a white belly. A raptor with black plumage, a white tail band and bright orange-red facial skin, the Black Caracara is

quite striking, and with its unique pattern is unlikely to be confused with any other species. As a forest dweller it is most often seen from boats travelling the rivers of the eastern lowland forest, and it often perches up in prominent places. As is usual with caracaras, it is an opportunist, being both a predator and scavenger.

LAUGHING FALCON *Herpetotheres cachinnans* 55cm

This characteristic bird of the lowlands is easily recognized by its cream-coloured underparts, dark brown back, a strongly patterned barred tail and a very distinctive, 'Lone Ranger'-style black mask. This is the most distinctive feature of this species apart from its voice, from which the bird's name derives – its presence is often given away by its loud 'laughing' calls. Sits up on prominent perches, particularly on dead trees, and is therefore rather obvious. It occurs in and around the edges of lowland forest east of the Andes, where it is fairly common in some places. A rather rare bird however on the Pacific slope of the forested Tumbes area in the north of the country. Occurs from Mexico to northern Argentina.

AMERICAN KESTREL *Falco sparverius* 27cm

This beautiful little falcon is possibly one of the first birds to be seen when landing at Lima's international airport. It is able to utilize all manner of open and even urban habitats, and is therefore one of the raptors most frequently encountered on the western slope of the Andes, especially along the coast. The male has a very striking pattern, a mix of rich rufous and grey above and whitish on the underparts, with a spotted pattern overlaid. The face pattern is also distinctive. Although common and widespread, this species repays a close look, as it is one of the most attractive members of the falcon family in South America.

BAT FALCON *Falco rufigularis* 25cm

A very attractive little falcon, and quite a common bird of the lowlands east of the Andes. Usually prefers areas close to water, so most often seen along the rivers and oxbow lakes of the region. Its plumage is very similar to the Orange-breasted Falcon *Falco deiroleucus*, but that species is much larger and much rarer. Bat Falcons are, as their English name suggests, partial to bats, and are therefore rather crepuscular in their habits, spending much of the day sitting quietly on dead trees. However, they are very fast in flight and will also hunt other birds, particularly swallows and martins. They are very dashing, both in appearance and style, and have the most beautiful plumage pattern, with a black head, white collar, rufous belly and black-barred breast.

SPECKLED CHACHALACA *Ortalis guttata* 53cm

Just one of a number of species in the family of Cracidae, this is the only common chachalaca in southern Peru. Most larger members of the family, such as the currasows, are rather rare, being large and conspicuous and so very often easily trapped or shot. This lowlandforest species is particularly loud and obvious. A rather plain brown bird, its most obvious plumage feature is the attractive speckled pattern on its breast. It also has a rather long and broad tail. Typical of the *Ortalis* genus, it moves around in bands, which often take part in very noisy sessions of 'chachalacking', sometimes annoyingly early in the morning! They will forage both on the ground and in trees, mostly for fruit, but will likely be looking for almost anything edible.

ANDEAN GUAN *Penelope montagnii* 60cm

Apart from the smaller chachalacas, this is perhaps the commonest member of the guan family in Peru. It is also the guan most regularly seen at higher altitudes in cloud forest, on the eastern Andean slope, and can be seen easily on the upper Manu road out of Cuzco. Largely frugivorous and so mostly observed feeding in fruiting trees, it is usually first detected by the rustling of the foliage of the trees, caused by the bird's ungainly manner of moving around. Although usually rather timid, the species can be quite tame and approachable in areas where not hunted. It has red wattles and striated body plumage. Occurs up to about 3500m.

HOATZIN *Opisthocomus hoazin* 63cm

One of the most extraordinary birds in the world, for a variety of reasons. First, it would be difficult to design such a strange-looking creature. Its habits and unique adaptations, which include being a ruminant, also set it apart from all other birds. A flamboyant 'punky' headdress, complex patterns on its wings, a long pale-tipped tail and strange calls, make it one of the most immediately recognizable riverside birds of the Amazon region. The young birds have hooks on their wings, which enable them to move around trees in a reptilian style, and they also possess the ability to dive underwater to escape predators. The species always nests over water. Hoatzins are rather smelly birds owing to their diet of semi-toxic leaves.

LIMPKIN *Aramus guarauna* 66cm

A rather strange large, long-necked and long-legged wading bird of the eastern lowlands, which requires wet habitats. Limpkins prefer areas with a lot of vegetation, and will sometimes perch in trees or skulk in bushy or reedy places, particularly in the heat of the day.

This can make them difficult to locate. They have a dark-spotted plumage and a long decurved bill to enable them to feed on water snails, which they extract from their shells. Can sometimes be seen from boats along vegetated riverbanks, but more usually on oxbow lakes or other backwaters. Most often seen in flight, when it resembles a large and somewhat ungainly ibis.

PALE-WINGED TRUMPETER *Psophia leucoptera* 55cm

A species of the lowland tropical forest, met with only occasionally. An unusual bird in that it is one of only three members of the family in the world, all of which occur only in the Amazon basin. Trumpeters are terrestrial and have a striking black-and-white plumage. Usually seen in groups along rainforest trails, as they forage through the undergrowth. They are so called because of their weird crooning and trilling calls, which are often heard at night. Can be quite tame but most definitely not in areas where they are hunted; they are sometimes captured to be kept as pets by indigenous people. Most easily seen in areas like the upper Rio Tambopata and Manu, where they are protected by the relatively low density of human habitation.

PLUMBEOUS RAIL *Pardirallus sanguinolentus* 27cm

Common but rather local, this bird is found in wet habitats in many parts of the country, from the coast to the high Andes. Quite conspicuous for a rail, as it spends a considerable amount of time foraging in open marshy habitats. Slaty-grey below and dark brown above, it has a slender pale-green bill with red and blue spots at the base, and red legs, making this a very handsome bird and rather elegant compared to its relative, the Common Moorhen *Gallinula chloropus*, with which it often shares its habitat. It is readily seen at Huacarpay lakes near Cuzco, and is replaced in the lowlands east of the Andes by the closely related and very similar Blackish Rail *Pardirallus nigricans*. That species has a darker bill lacking the red spot, and also a whitish throat, which this species does not have.

SPOTTED RAIL *Pardirallus maculatus* 25cm

Occurs in the northwest of Peru along the coastal strip south to La Libertad, where it is generally a rather uncommon bird, but this is probably largely due to the nature of the habitat it prefers, in which it survives largely unseen. It lives almost exclusively in dense marshland vegetation, reedy places and other wet areas, and appears to be particularly partial to rice fields. It will probably become more common as rice production in Peru increases. Possibly the prettiest of the South American rails, with its striking spotted and speckled plumage (which gives the bird its English name), barred underparts and red legs. It also has a greenish-yellow bill with a bright red spot at the base, which adds to its overall rather flashy appearance.

AMERICAN PURPLE GALLINULE *Porphyrula martinica* 33cm

An unmistakable member of the avian community of the marshlands, this beautifully coloured member of the rail family is found on oxbow lakes and marshes, mainly in the lowlands of the east, but has also been seen on the coast at the Villa marshes, just south of Lima. It is often conspicuous, especially early in the morning and in late evening, when it will be more likely to emerge from its normal reedy habitat to forage in the open. The immature birds resemble the closely-related Azure Gallinule *Porphyrula flavirostris*, but that species has a bluish neck, whereas young Purple Gallinules have a buff neck. It is a migrant, visiting the eastern lowland seasonally, and occurs widely throughout much of South America.

SUNGREBE *Heliornis fulica* 27cm

A member of the finfoot family, of which there are only three species. This is the smallest, and the only one in the New World. A secretive and attractive aquatic bird of the lowland rainforests, when swimming in the water it superficially resembles a grebe. However, it has a very different bill and there are no grebes that have a similar striped pattern on the head in adult plumage. Young grebes often have striped heads, but will normally be accompanied by adults. Apart from the head stripes, this species is rather plain brown above and white below. Usually stays close to banks of rivers and oxbow lakes and is often seen swimming under overhanging vegetation. Unlike grebes, this bird can walk on land, as is normal with the other finfoots, and is also able to climb up into trees.

SUNBITTERN *Eurypyga helias* 50cm

In a family of its own, this is one of the most amazing birds of Peru. An extraordinarily cryptically plumaged bird, it has a fantastic display to deter predators. This involves opening the wings and flashing eye-like markings, which makes it look like more like a giant moth than a bird. Its complicated plumage pattern also allows it to remain inconspicuous in open environments, blending easily with its surroundings, especially when on a stony river beach lacking any vegetation, as shown in the photograph. Lives mostly along streams and oxbow lakes of the eastern lowlands and lower Andean slopes, where it feeds on various aquatic creatures. A truly enigmatic and spectacular bird.

WATTLED JACANA *Jacana jacana* 25cm

Also known as the 'lily trotter', owing to its ability to walk on aquatic vegetation, including lily pads. It does this by utilizing its immensely long toes, which spread the bird's weight. It also has long thin legs. A well-named bird, as it has large red facial wattles! It has rich deep chestnut upper-parts and blackish body plumage, but perhaps the most extraordinary distinguishing feature are the startling yellow wing feathers, which are apparent only when it takes flight. The immatures are white below. Quite a common bird in the lowlands east of the Andes in its wetland habitat, which is mainly oxbow lakes with grassy margins.

BLACKISH OYSTERCATCHER *Haematopus ater* 48cm

As its name indicates, this large and stocky shorebird has a totally dark plumage, brownish on the upperparts and charcoal greyish-black below. It is a common sight along the rocky coasts of southern Peru. Here it forages largely for mussels, opening them with its large and powerful bright orange bill, which is its most obvious feature since it is otherwise rather camouflaged. Has a call that is very similar to the Eurasian Oystercatcher *Haematopus ostralegus*, but obviously has a completely different appearance from both that species and the black-and-white American Oystercatcher *Haematopus palliatus*. It also has very pale, whitish-pink and extremely thick and sturdy legs.

AMERICAN OYSTERCATCHER *Haematopus palliatus* 43cm

This oystercatcher is also a common and obvious feature of the Peruvian coastline, but looks totally different from its Blackish *Haematopus ater* relative. It is strikingly patterned, with a dark brown back, a somewhat contrasting black head and white belly, and an obvious white wing-bar. It is also a somewhat smaller and slighter bird than its congener and has a rather different voice. It does, however, share its close relative's most distinctive feature, which is the large and powerful bill that is used to probe for worms, or at other times to open the shells of mussels.

PERUVIAN THICK-KNEE *Burhinus superciliaris* 40cm

This species is one of nine members of the global family known as stone curlews. Typically crepuscular and nocturnal, it roams widely at night to forage in marshy areas, rough grazing land and other agricultural areas along the coastal plain. The huge white eyes are evidence of this ability. Spends much of the day in low profile, waiting quietly for the sun to set. It has long legs but is rather anonymous in general appearance, except for a very obvious supercilium and a prominent black stripe above it. Makes far-carrying and rather mysterious-sounding wailing calls, usually at night.

GREY-BREASTED SEEDSNIPE *Thinocorus orbignyianus* 23cm

One of four members of the family, three of which occur in Peru. Common in the puna zone, mainly in the south of the country, this species occurs all the way up to the snow line at about 5000m. The male has the grey breast, whereas in the female this is buff. It also has an exquisitely complex and beautiful patterning on its upperparts, which is apparent when seen at close range. Often flushed from the sides of roads running through the puna grasslands, where it likes to forage along the verges, but is also likely to be seen when walking through its high grassland habitat.

SOUTHERN LAPWING *Vanellus chilensis* 33cm

This large and handsome lapwing prefers grassy habitats, and with deforestation is likely to become a more common bird in the southeast of the country and around Iquitos in the north. Strangely, as it is a common bird in open habitats in many other parts of South America, it does not occur on the coast. It is a beautiful and ornate plover, with a crest and a rather unusual red and black pattern on its legs. As is normal with the *Vanellus* genus of plovers, it is also a noisy and aggressive species, often alerting other birds with its loud calls to a predator, or to people who are approaching the area where a nest is situated. It will attack them directly if they stray too close.

PIED LAPWING *Vanellus cayanus* 23cm

This species is rather dubiously placed in the genus *Vanellus*, as it is rather different from the other lapwing species. It is a considerably smaller species, and rather different in appearance, with its lovely black-and-white pattern. Quite a common bird along the banks of rivers that flow through the lowland forests. The attractive plumage is very striking but actually rather cryptic, and on the open riverbanks

on which the bird lives its plumage pattern is clearly designed to protect it from predators. Although quite common, its pattern and behaviour can sometimes make it surprisingly difficult to spot, unless it is running on an open sandbank. On stony beaches, too, it can be very hard to pick out. Most often seen from boats plying rivers such as the Madre de Dios, along which there are many sandy and rocky beaches.

ANDEAN LAPWING *Vanellus resplendens* 33cm

One of the most conspicuous and noisy birds of the high Andes, often found in irrigated areas close to human habitation, but also in more remote marshy or damp grassy areas. It is a rather common bird, and has a very distinctive, mostly green and grey plumage pattern. A very typical member of the *Vanellus* family, with its loud alarm calls and striking plumage. It also has vivid red eyes. As its name suggests, it is common at high altitudes, occurring in suitable habitat up to 4500m. However, as with many high Andean birds in Peru, it will very occasionally come down to the coast, especially when conditions are particularly severe at high elevations.

SEMIPALMATED PLOVER *Charadrius semipalmatus* 17cm

This species and the Snowy Plover *Charadrius alexandrinus* are closely related, but this bird is a migrant from North America, whereas the Snowy is present at all times of the year. Breeds mainly in coastal areas around the edge of the Canadian tundra. It is characterized by its white collar and black breast-band, and is a much stockier species than the Snowy Plover. It is also a darker shade of brown on its upperparts than its relative, and has orange legs. It has been recorded at high altitude on occasion in Peru, but is nearly always seen on the coast, normally between November and April, where it is a commonly occurring member of the plover family during those months.

SNOWY PLOVER *Charadrius alexandrinus* 15cm

A common bird along the beaches of the coast throughout the year, although there is undoubtedly some local migration and probably an influx of birds from the north during the boreal winter as well. Characterized by its incomplete breast band and more pallid upperparts compared to its long-distant migrant cousin, the Semipalmated Plover *Charadrius semipalmatus*. Outside the breeding season both sexes are far less well marked, but during breeding the male has a striking pattern of black marks on each side of its neck. This species also has dark legs as opposed to the orange legs of its close cousin. Forages along the beaches for tiny invertebrates.

KILLDEER *Charadrius vociferus* 25cm

A member of the plover family, and common in a variety of lowland areas, especially grasslands close to water and agricultural areas near the coast. Distinguished by its double breast-band and its long wings and tail, the bird has an elegant and elongated profile. It also has a very pretty chestnut tone to the upperparts and a rufous rump. A noisy and conspicuous bird, its English name is onomatopaeic, based on its high-pitched call 'kee-dee, kee-dee, kee-dee'. As with many other species, the call is often the first indication of the bird's presence, particularly at night.

DIADEMED SANDPIPER-PLOVER *Phegornis mitchellii* 18cm

One of the most enigmatic birds of the most remote and sometimes barely accessible areas of the very high Andes. Uncommon and very local, depending on the availability of its unique habitat. Breeds in fluvial bogs, which are often frozen at dawn on a daily basis, or even covered in snow. Its beautiful and cryptic plumage is unique, with finely barred underparts, dark head and rufous collar. Can be most readily seen in the bog near the road to Milloc, which is only a few hours' drive from Lima. But people prone to altitude sickness should be aware that driving from Lima to Milloc at 4000m altitude may make them feel rather ill. You will also need to take waterproof footwear to walk in the bog if you wish to see this remarkably plumaged bird!

SOUTH AMERICAN SNIPE *Gallinago paraguaiae* 27cm

A generally uncommon bird of the lowland wet grasslands and marginal habitats. It is a very typical member of the family and closely resembles another, namely the North American Wilson's Snipe *Gallinago (gallinago) delicata*, which may occasionally arrive from the north as a migrant. It is currently almost impossible to separate the two in the field, but they do have slightly different calls. As the photograph illustrates, this bird can, with care, be approached closely, as it often sits very tight until finally putting up and uttering a typical short harsh call. Clearly this bird is typified, as is shown in the photograph, by its cryptic plumage and in particular its very long bill, which it uses to probe damp ground or mud for its food.

LESSER YELLOWLEGS *Tringa flavipes* 25cm

A common migrant to Peru from its Arctic breeding grounds, this bird occurs in many wetland areas during the northern winter months, between October and March. The smaller cousin of the Greater Yellowlegs *Tringa melanoleuca*, this is a much slimmer and more delicate bird, although in terms of their plumage the two species are very similar in general appearance. However, that displayed by the Lesser has a much more spangled and finer pattern. The photograph shows very graphically the difference between the two species, the larger Greater Yellowlegs (on the right) having much sturdier legs and a heavier bill than its smaller relative. In common with its larger cousin, the Lesser shows a white rump and trailing yellow legs when in flight.

GREATER YELLOWLEGS *Tringa melanoleuca* 35cm

This common migrant from the far north is seen regularly during the northern winter months on both sides of the Andes, where its loud '*tu tu tu*' call is commonly heard during this time. Rather similar in its general plumage pattern to its smaller cousin, the Lesser Yellowlegs *Tringa flavipes*, and clearly in also having yellow legs. However, this bird

is considerably larger and has a darker and coarser plumage pattern. The smaller Lesser Yellowlegs also has a higher-pitched call. Often found on coastal lagoons during its stay in the country, but tends to have less of a particular preference than its relative for fresh-water habitats. It can be seen in saline habitats along the coast as well as along the rivers of the eastern lowlands, and also commonly at high altitudes.

WILLET *Catoptrophorus semipalmatus* 40cm

Quite similar to, but considerably larger than, the Greater Yellow-legs *Tringa melanoleuca*. However, it is clearly lacking that bird's most obvious feature by having dark legs, and it also has a much heavier bill. Easily recognizable in flight by the very obvious whitish patches at the base of the flight feathers. It is a fairly common migrant to coastal Peru during the boreal winter months, when it arrives from its breeding area in North America as far north as Canada. Has also been recorded at high altitudes, although this a rare occurrence, and is usually to be found running or probing on sandy shores or foraging along more rocky areas.

BLACK-NECKED STILT *Himantopus mexicanus* 50cm

A common species in South America, this bird has extraordinarily long, bright pink legs. Its long, straight bill and very obvious black-and-white pattern also make it instantly recognizable as a stilt. It is rather similar to the White-backed Stilt *Himantopus melanurus*, but that species has a white crown and collar and also tends to be restricted to the lowlands east of the Andes. The Black-necked Stilt can be seen in many parts of Peru, being common in wetland habitats along the coast as well as in the high Andes at over 4000m, at Lakes Junin and Titicaca. A noisy and sometimes aggressive species, which will attack people or animals that approach its nest.

RUDDY TURNSTONE *Arenaria interpres* 23cm

Although it breeds on the Arctic tundra, the Ruddy Turnstone visits Peru during the northern winter and can be found anywhere along the coast, sometimes in the most unlikely places. As its English name suggests, this species is generally found foraging on rocky shores, turning over stones and seaweed in search of prey. However, unlike most migrant shorebirds, Turnstones are exceptionally opportunistic. This photo was taken in a beachside restaurant where the birds were feeding on peanuts, sometimes even on the bar top! I have also seen this species feeding on discarded sandwiches and even scavenging on other dead birds.

WHITE-RUMPED SANDPIPER *Calidris fuscicollis* 18cm

Unlike the Baird's Sandpiper *Calidris bairdii*, this species is a rather uncommon migrant from the Arctic. The two species are rarely seen together, since they do not follow the same migration routes, Baird's mainly using the high Andean flyway. However, the White-rumped does occur on both sides of the Andes and is occasionally seen both on the coast and along the lowland eastern rivers on its journey to the southern shores of Patagonia. A long-winged and elegant sandpiper, it can be distinguished from the Baird's Sandpiper by its combination of a white rump, which is very apparent in flight, and its overall greyer plumage.

BAIRD'S SANDPIPER *Calidris bairdii* 18cm

A bird that comes through Peru in large numbers, heading south from its North American breeding areas in September and back north again in April. However, unlike the other members of the family, this is a bird most often seen in the highlands. It winters further south and has one of the longest migrations of all American birds. Often seen at Huacarpay lakes near Cuzco. It is a long-winged and rather elegant member of the *Calidris* sandpiper family, somewhat similar to the White-rumped Sandpiper *Calidris fuscicollis* but generally a brighter, more golden-hued bird. Can sometimes be very tame and approached very closely.

CHILEAN SKUA *Stercorarius chilensis* 58cm

One of a pair of large and some-times rather similar looking gull-like predatory seabirds that occur in Peruvian waters. This is a very heavy and robust species and always has rather rufous or cinnamon tones on the plumage of its underparts and around the neck. Has streaky brown upperparts and broad, pointed wings with very obvious white patches at the base of the flight feathers, a typical feature of the skua family. Predates other seabirds, especially their young, and attacks them in order to force them to disgorge their stomach contents. Breeds on the islands of southern Patagonia and occurs in the region during the non-breeding season.

SOUTH POLAR SKUA *Stercorarius maccormicki* 52cm

This species of skua is rather lighter and more agile than the Chilean Skua *Stercorarius chilensis*. It also has a completely different lifestyle, being a very long-distance migrant; it passes through the waters of Peru on its long, southward journey from the north Pacific as it heads back to its breeding grounds on the Antarctic Peninsula. Normally occurs in October or November but, unless adverse conditions force it closer to the coast, it will only be seen from a boat, far offshore. This skua has a variety of plumage types; some have very light underparts whilst others are quite dark, but they are always much plainer than the Chilean Skua, always lacking the cinnamon tones of that species, albeit sharing white wing flashes.

SWALLOW-TAILED GULL *Creagrus furcatus* 50cm

Another seabird which visits Peruvian waters from the Galapagos Islands, where it nests. A rather large and quite strange-looking gull, with a very striking wing pattern of large white patches and mostly black outer flight feathers. It also has a quite obviously forked tail, which in the immature has a black band at the tip. It has very large eyes, a clue to its mainly nocturnal lifestyle, by which it forages at sea during the night. In the non-breeding season this gull can be seen off the coast as far south as Chilean waters, but always offshore and never coming to land during the time it spends in Peru. The photograph shows an immature bird.

GREY GULL *Larus modestus* 44cm

This attractive and rather delicate-looking gull is one of the most distinctive gulls of the region and quite a common bird on the beaches of Peru. The adult (illustrated) has a dove-grey body and a whitish head in breeding plumage; in the non-breeding season it has a dark head. The species is also characterized by its long-winged and streamlined profile. It has quite a long and slender black bill and a white trailing edge to the wing. The immature is browner and has a dark brown head. Breeds inland in the southern Atacama Desert, and is a migrant that moves in and out of Peru seasonally from the south.

BAND-TAILED GULL *Larus belcheri* 50cm

This handsome gull is easily recognized by its black tail-band and is one of the commoner gulls along the Peruvian coast, found as far north as Piura. It is superficially similar to the larger Kelp Gull *Larus dominicanus*, the adults having blackish upperparts. It is nearly the size of the Kelp Gull, but its larger relative lacks the diagnostic black tail-band and has a much heavier bill with a different pattern, which is typically yellow with a red spot on the lower mandible. In this species the bill has a completely red tip and black band. Young birds and adults in non-breeding plumage have blackish heads.

KELP GULL *Larus dominicanus* 53cm

Adult   Young

This is the common, large, black-backed gull along the coast. It is rather larger than Band-tailed Gull *Larus belcheri* and lacks that species's black tail-band in adult plumage. It does, however, have a dark tail-band in its many guises during the four years it takes to mature. As is typical in gulls, its plumage gradually lightens with each successive moult until it attains its crisp black-and-white adult dress. It can always be identified by its large size. A powerful and predatory bird, it will commonly chase and sometimes catch and kill other birds, or raid the nests of birds such as the South American Tern *Sterna hirundinacea*. In the non-breeding season it is largely a scavenger, or forages along the shoreline.

FRANKLIN'S GULL *Larus pipixcan* 35cm

During the northern winter months this pretty gull arrives in huge numbers and is extremely common during this period. It migrates from the Canadian prairies and follows the Pacific coast, arriving in Peruvian waters in early November. Almost the entire world population arrives in or passes through Peru, and tens of thousands may occur along the coast in the Paracas area in January and February. During the months that it remains in Peru, it will usually be in winter dress, and therefore lacking the black head of its breeding plumage. It is characterized in adult plumage by the obvious black wing-tips with contrasting white spots, which are known in gulls as 'mirrors'.

SOUTH AMERICAN TERN *Sterna hirundinacea* 43cm

This large tern is fairly common along the coast, and can be found nesting along the edge of the cliffs on the Paracas peninsula, for example. Like most terns, it will attack people if its nest is approached too closely. A very elegant bird with a long tail and similarly long wings, it has a light and buoyant flight. The black cap and long red bill also set it apart from gulls and are very typical features of the genus *Sterna* when in breeding plumage. In the non-breeding season it loses the black crown, which becomes mostly white, and the bright red bill becomes darker. Like other terns, it is susceptible to predation by gulls whilst nesting, and by migrant skuas (jaegers) when at sea.

INCA TERN *Larosterna inca* 40cm

One of the most easily recognized of Peruvian seabirds, this very beautiful tern is mostly grey, with long down-curling cream-coloured plumes on the sides of the head, and a red bill and legs. The young birds, however, are rather less obvious and appear uniformly rather dark grey, but are clearly smaller and more elegant than the Grey Gulls *Larus modestus* with which they are often seen. Common along the coast and very easily seen from the shore in Lima. Breeds in quite large numbers on the offshore islands, the Ballestas in particular. However, it can very often be seen from the mainland coast, fishing just offshore, or perched on the jetties along the beach in Lima itself.

LARGE-BILLED TERN *Phaetusa simplex* 38cm

This large tern is very distinctive with its massive yellow bill, and is an obvious, common and easily recognized bird of the Amazonian lowland river systems and oxbow lakes. Unlike most terns, it does not usually occur in coastal waters, being normally a freshwater species. Often found nesting on sandbanks alongside Black Skimmers *Rynchops niger*, it has a very striking pattern on the upper wings, which consists of three triangular areas of black, white and grey. Commonly follows ships travelling along these lowland rivers, which churn up prey from the normally muddy and murky waters of the rivers that flow off the Andes.

BLACK SKIMMER *Rynchops niger* 45cm

This extraordinary bird is common along the lowland rivers east of the Andes, and along the coast in the northern winter months. One of only three species in the world which feed by flying just above the surface of the river with their lower mandible in the water and snapping up prey items as they go along (*inset*). Skimmers have the unique adaptation of having the lower mandible considerably longer than the upper one to enable them to perform this spectacular feeding feat. One of the most easily recognized birds along the rivers, with its strange red and black bill and its black-and-white plumage, with prominent white trailing edge to the long, pointed wings.

PACIFIC DOVE *Zenaida meloda* 27cm

One of the commonest birds in the Peruvian capital city, Lima, where this species is very often one of the most obvious birds in the urban streets and is frequently seen commuting across the city in flocks from one favoured feeding place to another. Its cooing call competes for attention with the song of Rufous-collared Sparrows *Zonotrichia capensis*. It also occurs widely in the countryside. Although it has a rather plain appearance, being mainly a mid-brown colour, it is characterized by the white bands on its forewings and at close range the purple-blue skin patch around its eyes is apparent and rather attractive.

EARED DOVE *Zenaida auriculata* 25cm

One of the most widespread doves of the region. Fast-flying and rather anonymous until seen well, when the subtle but attractive plumage pattern is apparent. This consists of various shades of brown with dark spotting on the wings, pinkish hues on the underparts and a white vent. At close range, however, it is more distinctive and displays a pretty, iridescent purple-pink neck patch and the small dark mark on the sides of the head which gives the bird its name. Common in both highland and lowland areas. Although it superficially resembles several other dove species in having white outer tips to the tail, this species has a distinctive pointed tail.

69

RUDDY GROUND-DOVE *Columbina talpacoti* 18cm

One of the most attractive members of the ground dove tribe, of which there of are ten species in Peru. This is a common and obvious member of the avian community in the Amazonian lowlands, often flushed from roadsides and found at the edges of villages, in gardens and various other open or marginal habitats. It is quite a tame bird and so is often seen well. The male is distinctive, having rusty-coloured upperparts, which contrast with a soft blue-grey cap and nape and a paler and pinker underside. The female has less obvious reddish hues and somewhat resembles the Plain-breasted Ground-Dove *Columbina minuta*. However, females are nearly always accompanied by males, giving a strong clue to their identity.

PALE-VENTED PIGEON *Columba cayennensis* 30cm

This lovely and delicate pigeon of the lowland forests is quite a common bird, often seen sitting high up on exposed perches in the early morning or evening, when its maroon-red upperparts, pinkish breast, white throat and grey rump and tail are apparent. As its English name suggests, it has a contrasting white vent, but this is not always obvious. Perhaps most often seen flying across rivers. It utters a soft, four-note hooting call. Could be confused with the Ruddy Pigeon *Columba subvinacea* in certain light conditions, but that species is always much more uniformly coloured.

BLUE-AND-YELLOW MACAW *Ara ararauna* 83cm

One of the largest and most instantly recognizable of the parrots, with beautiful plumage and a long tail. Still common in areas of intact lowland forest east of the Andes, this spectacular bird is most often seen crossing rivers in pairs or threes as they commute from one place to another. One of the three will be a youngster. Like most members of the parrot family, it routinely attends the gatherings at clay licks on riverbanks, where it will dig out and swallow the mineral-rich earth. This is assumed to help in the digestion of certain fruits, but is also clearly a social event, with the birds spending much time interacting.

RED-AND-GREEN MACAW *Ara chloroptera* 96cm

Another large and beautiful member of the parrot family. Rather similar to the Scarlet Macaw *Ara macao*, but is darker red and can be separated from that species by the lack of yellow patches on its wings. It also displays prominent facial striping, actually lines of tiny red feathers. Can be seen in large numbers at Blanquillo on the upper Rio Madre de Dios, where they gather to socialize and eat minerals at the clay-lick. They usually exclude all other macaws from these gatherings, but will sometimes allow one or two Scarlets to join in. As with other macaws, it has a loud and raucous call.

YELLOW-FACED PARROTLET *Forpus xanthops* 14cm

One of the four *Forpus* parrotlets occurring in Peru, this tiny member of the parrot family is the rarest and is endemic to Peru, occurring in the semi-arid areas in the valley of the Rio Marañon. It is a very localized bird of the north of the country, but in some particular localities can be quite common. Its bright yellow face is diagnostic. As with the other parrotlets, it is very short tailed. Can be seen in groups in flowering or fruiting trees, where they gather in noisy congregations. Perhaps most regularly seen along the road from Leimabamba to Balsas, close to the bottom of the deep Marañon valley, although it sometimes occurs at much higher altitudes (up to 2750m). These movements are probably seasonal and depend on food sources from fruiting trees.

MITRED PARAKEET *Aratinga mitrata* 38cm

This moderately long-tailed *Aratinga* is just one of several species which all look very similar, in particular the Scarlet-fronted Parakeet *Aratinga wagleri* and the White-eyed Parakeet *Aratinga leucophthalmus*. This is the only one of the three that does not have any red on the bend of the wing. However, it has variable amounts of scattered and random red markings on the sides of the neck and on the nape. Found in the highlands from about 1000m, sometimes up to 4000m, and roams around seasonally in large, noisy flocks. It can often be seen doing just this along the road from Cuzco to Manu at about 3500m elevation.

MEALY PARROT *Amazona farinosa* 40cm

This is the largest of the Amazon parrots and a charming member of the parrot family, as it spends much time in apparent conversation with other Mealies. Its loud and lovely gurgling voice is very distinctive and far-carrying, and sounds as if it has a diverse and quite complicated vocabulary. It is rather common in the lowlands of the east, where there is still good forest habitat, but its nests are often raided by local people for the young, as they grow up to make good pets. It is largely green above and below, with a greyish-lavender nape and a large white eye-ring. Shows a striking tail pattern with a broad yellow band. The name Mealy refers to the variably pallid and somewhat greyish tone of the bird's upperparts.

LITTLE CUCKOO *Piaya minuta* 27cm

As its name suggests, this is one of the smallest members of the *Piaya* cuckoos and is typical of the group in having a rich rufous plumage. It has a much shorter tail than its cousin, the Squirrel Cuckoo *Piaya cyana*, but shares the white tail-tips of its larger congener. It is an inhabitant of the eastern humid lowlands and is often found in quite marginal habitats, such as forest edge and dense secondary growth, often close to water. It is more often heard than seen. This is also a much less common species in Peru than its larger and longer-tailed relative, despite not being a forest inhabitant.

SQUIRREL CUCKOO *Piaya cayana* 40cm

This species is the commonest member of this group of cuckoos. Very often seen in the lowland forests of the east and also on the western Andean slope in the north, where there is forest. It is a large and obvious bird, and with its long tail with obvious white tips, bright rufous plumage and conspicuous yellow bill, it is hard to miss as it crashes around in the trees hunting for caterpillars and other prey. It also has a bright red orbital ring around the eye. Its call might be transcribed as 'Kick yooooooou', the first note having an upward inflection and the second note downward.

GREATER ANI *Crotophaga major* 48cm

Typically a bird of the thick river-bank foliage along the streams and rivers of the eastern lowland forests. At first glance appears to be a rather large and ungainly black bird with a long tail, but in bright conditions can be seen to be beautifully iridescent, showing blue and green reflections off its upperparts and purple and violet reflections off the tail. Usually seen moving in groups through dense riverside vegetation or flying across steams and rivers. This is the largest of the anis in Peru, as its name suggests. The other two species, the Groove-billed Ani *Crotophaga sulcirostris* of the west coast and the Smooth-billed Ani *Crotophaga ani* of the eastern lowlands, are both common within their range but are much smaller and dowdier than this species.

BURROWING OWL *Athene cunicularia* 23cm

The most obvious owl of the coastal areas and the only one that is strictly terrestrial, preferring very open areas which are sometimes almost devoid of vegetation. It nests and spends much time in underground burrows. However, when trying to spot potential prey it will perch up on fence-posts or other vantage points in a very conspicuous fashion. At other times it tends to remain on the ground close to its burrow. This can either be an abandoned mammal burrow, or one that the owl has excavated itself. When on the ground its cryptic plumage can make it difficult to spot, but it can be quite tame, especially near urban areas, where it is familiar with people.

FERRUGINOUS PYGMY OWL *Glaucidium brasilianum* 16.5cm

The commonest member of the *Glaucidium* group in the humid eastern lowlands, and the lower foothills on the eastern side of the Andes. It is perhaps conspecific with, and at least very closely related to, the Peruvian Pygmy Owl *Glaucidium peruanum*, but was recently deemed to be a separate species. Its plumage is very variable, but usually has a rufous tone. The crown is usually finely streaked, but can appear almost plain. It is replaced at higher elevations along the Andean slope by the Andean Pygmy Owl *Glaucidium jardinii* in the north and Yungas Pygmy Owl *Glaucidium bolivianum* in the south. Both of these species have spotted crowns.

PERUVIAN PYGMY OWL *Glaucidium peruanum* 18cm

As is normal with the pygmy owls, of which there are six species in Peru, this bird closely resembles several others. As the photograph illustrates, it is a dry-country bird, which likes the intermontane valleys at medium altitudes, often perching on cacti. For example, it can be seen in the valleys of the Marañon river system, Cuzco, and it is also the only pygmy owl occurring on the west slope of the Andes and the coastal plain. It is very similar to the Ferruginous Pygmy Owl *Glaucidium brasilianum*, which is much commoner on the eastern side of the Andes at lower elevations and in forest environments where this species does not occur. Both species have spotted mantles and streaked crowns.

OILBIRD *Steatornis caripensis* 45cm

A very strange and rather mysterious nocturnal creature and the only strictly frugivorous night-flying bird. Lives in caves by day, often in quite large numbers, especially the large colony at the Cueva de las Lechuzas, near Tingo Maria in Huanuco province. It emerges at dusk, often flying long distances to find fruiting trees. It occurs in the lowlands and foothills of the east, where there are still tracts of forest in which it can successfully forage for fruit, and where there are caves in which it can roost and nest. However, it is only found very locally in a few areas. It is a rusty-brown bird with long wings and tail, with white spots on both.

COMMON POTOO *Nyctibius griseus* 38cm

This species is also sometimes known as the Grey Potoo. It is usually found in lowland forested areas, although seems to prefer more open habitats than some of its larger relatives, such as the Great Potoo *Nyctibius grandis* and the Long-tailed Potoo *Nyctibius aethereus*. In most places this is the commonest of the six species of this cryptic, nocturnal group of birds to occur in Peru. They are strictly insectivorous and spend the day in a motionless position, disguised as part of the tree in which they are perched. This species has a beautiful but mournful call, which consists of four or five descending notes and is heard as dusk descends.

LESSER NIGHTHAWK *Chordeiles acutipennis* 20cm

Members of the nightjar family, the nighthawks are characterized by their long, pointed wings, which give them a somewhat falcon-like shape. Combined with their fast and rather aerobatic flight, this sets them apart from most of the other nightjar species, of which there are twenty in Peru. Seven of them are nighthawks and this species is the smallest. It resembles rather closely the Common Nighthawk *Chordeiles minor*, which is a migrant to the region but only a rare visitor to the north of the country. This species occurs on both sides of the Andes in open areas, where it can be seen hawking for insects at dawn and dusk.

PAURAQUE *Nyctidromus albicollis* 26–28cm

One of the commonest members of the nightjar family in lowland forested regions. Frequently occurs in marginal habitats, including cleared areas close to human habitation and secondary growth. It is not generally found inside primary forest, but often sits on tracks and roads running through forested areas, situations in which it is most often and most easily seen. Nearly always the first indication of its presence is its distinctive and somewhat ethereal wailing call, which may be transcribed as a lilting '*koowheeeo*'; a very distinctive and familiar sound of the lowland forest. It has a lovely plumage pattern, having rufous 'cheeks' and a series of very bold markings on the upperparts. The male flashes white in the tail.

LADDER-TAILED NIGHTJAR *Hydropsalis climacocerca* 25–26cm

A lowland species, this nightjar is peculiar in that it is closely associated with water and therefore most often seen hawking over rivers or perched on riverside features such as branches protruding from the water. Quite common in the Amazonian region. Named for its rather intricate tail pattern, with white inner webs and a series of black bars. The male has a somewhat longer and more strongly patterned tail than the female. The body plumage of both sexes is rather pale, the male being greyer with a white band on the wing and the female more buff-coloured, with a correspondingly buff wing-band. Can be seen readily along the Rio Madre de Dios at Manu Wildlife Centre.

LYRE-TAILED NIGHTJAR *Uropsalis lyra* ♂ 75–80cm, ♀ 25cm

A beautiful nightjar whose most obvious feature is the male's extraordinarily long double-pronged tail. It occurs at middle elevations in forest on the eastern slope of the Andes, and can be seen from the bridge over which the Manu road from Cuzco crosses the Rio Union at about 1800m. Here the males can be watched as they display over the river at dusk, showing off their fabulous tails. The female is a much more anonymous creature, lacking these wonderful tail adornments. Both sexes have a dark, cryptic plumage and a rusty collar around the back of the neck. Occurs between 800m and 3500m and regularly nests along the rocky sides of road cuts.

SCISSOR-TAILED NIGHTJAR *Hydropsalis torquata* ♂ 40–50cm, ♀ 23–26cm

Another long-tailed species, but unlike the Lyre-tailed Nightjar *Uropsalis lyra* this is a lowland bird found in the south and east, usually along forest edge and particularly on dry river beds. Can be seen in Pampas del Heath in the extreme southeast of the country. The male has long tail streamers, which are dark with prominent white shafts, making him easily recognizable. As is typical in this group of nightjars, the female lacks the spectacular tail and her appearance is much more anonymous. Probably most often seen in car headlights at night, as it regularly enjoys sitting on roads. Has a lovely rising and falling call. It is a rather uncommon bird and, although it closely resembles Swallow-tailed Nightjar *Uropsalis segmentata*, the two never occur together since that species is a high-altitude dweller.

SHORT-TAILED SWIFT *Chaetura brachyura* 11cm

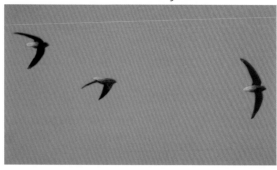

This is a common little swift of the eastern Amazonian lowlands and possibly the one most often seen, usually flying very fast and erratically over the forest. One of an amazing total of fifteen swift species in the country, six of which belong to the genus *Chaetura*. This genus contains many very similar species which are notoriously difficult to identify. As the name suggests, the tail is very short, giving this bird a distinctive profile and making it the easiest to identify. It has an extremely rapid bat-like flight and rather short and paddle-shaped wings for a swift. Also has a pale, caramel-coloured rump and upper tail.

RUFOUS-BREASTED HERMIT *Glaucis hirsuta* 10.5cm

This large member of the hermit group of hummingbirds is quite common in the eastern forested lowlands, and unusual in that it displays neither the elongated white central tail feathers that typify most of the hermits nor the characteristic black-and-white face pattern of the *Phaethornis* group. It lives in the forest undergrowth, feeding from widely scattered flowers and covering many miles each day. It has a long decurved bill, which it uses to pick insects from the underside of leaves as well as probing deep into flowers. Its large size allows it to dominate other hummers at bird feeders.

REDDISH HERMIT *Phaethornis ruber* 7.5cm

One of about a dozen species of hermit in Peru. This group of hummers is interesting for the fact that they are lowland birds, whereas most other hummingbirds live in the highlands. They also have a different feeding strategy, in that they have a large territory in which they visit many different but specific flowers in succession and on which they are reliant. This is commonly known as 'traplining'. If one wears a brightly-coloured (normally red) lapel pin or badge on a hat, a hermit will often hover in your face, checking out the prospects of a sugary meal, before zooming off to find a real flower. Reddish Hermits perform an amazing aerial display, in which the male bounces up and down over the female like a yo-yo, as shown in the photograph.

GREY-BELLIED COMET *Taphrolesbia griseiventris* 15cm

A very rare and localized bird, occurring only in a small area in the arid north of the country near Cajamarca. Lives in narrow gullies, where the drainage encourages a relative profusion of flowers, creating a more benign environment for a hummingbird amid the generally rather inhospitable surroundings. Sits up high and obvious, so where it is present it is easily seen. This means that this hummingbird is clearly apparent in the few places where it occurs and can therefore truly be known to be a rare creature. It somewhat resembles a Long-tailed Sylph *Aglaiocercus kingi*, having quite a long and deeply forked tail, but is a considerably larger bird and visually does not compete with the long, violet and purple streamers of that species.

MARVELLOUS SPATULETAIL *Loddigesia mirabilis* 12.5cm

Possibly the most remarkable of Peru's myriad of hummers, and a victim of its own strangeness and beauty; the heart of the male is considered to be an aphrodisiac! To save this 'marvellous' little bird, a concerted effort is underway to persuade local men that this is not so. Inhabits a very limited range on only one side of the

Utcubamba river valley, in the north of the country. Often, when glimpsed in flight, the most obvious feature of this species is not the bird itself but the strangely formed and extravagant tail feathers that give the bird its name. These feathers are wire-like with large, shiny black 'spatulas' on the end, and are used by the males in an extraordinary display, when they are lifted high overhead whilst in hovering flight.

SWORD-BILLED HUMMINGBIRD *Ensifera ensifera* 13cm

One of the most extraordinary birds in the world, having a bill as long as, or even longer than, its body, so although its length is supposedly about 13cm, much of this is the bill! This amazing adaptation enables it to feed from long trumpet-shaped flowers such as Daturas. Occurs on the east Andean slope at typical elevations of between about

2000m and 3000m, where there are abundant shrubs with these typically tube-like hanging flowers. It is of course instantly recognizable, but is a rather uncommon bird. Seen with some regularity along the Manu road out of Cuzco at about 2600m elevation, and at the same altitude along the road from Ollantaytambo to Quillabamba, below the east side of the Abra Malaga pass.

WHITE-NECKED JACOBIN *Florisuga mellivora* 11–12cm

A hummingbird of the lower slopes of the Andes and the forest of the Amazon lowlands. Characterized by its white underparts and the gleaming white hind collar. This lovely hummingbird is commonly seen at feeders, where it will compete viciously with other species to protect this unnatural food source. It often fights, particularly with Banana-quits *Coereba flaveola*, for access to 'its' feeder. Will at other times feed on a variety of flowering trees, where it has to compete seasonally with the large and aggressive Sparkling Violetear *Colibri coruscans*. It has a violet-blue breast and head which, with its deep green back and white collar and belly, make a very striking combination.

BLACK-TAILED TRAINBEARER *Lesbia victoriae* ♂ 24cm, ♀ 14cm

This is one of the commonest hummingbirds of the highlands. The male is recognized by its long black tail, but the female is, as is usual with hummingbirds, very different, and lacks the long tail. She has some spotting on pale underparts with a buffish throat and breast. This species has been adept at utilizing the many introduced Eucalyptus trees, when in flower, and will defend a flowering tree with quite some force. The very similar Green-tailed Trainbearer *Lesbia nuna* is the closest relative, but its tail is shorter, and of course green. Can readily be seen around Cuzco where there are many Eucalypts, especially if one were to walk up to the ruins of Sacsayhuaman above the town.

SPARKLING VIOLETEAR *Colibri coruscans* 13cm

This large and rather common hummingbird is one of the most regularly encountered species in the highlands. At certain times of the year it will migrate altitudinally to take advantage of flowering lowland forest trees, where it gathers in groups of the same species and will defend a whole tree from other hummingbird species, often with some viciousness. At high altitudes it is one of the most ubiquitous hummingbird species, being very catholic in its habitat preferences and decidedly opportunistic. Very similar to the Green Violetear *Colibri thalassinus*, but characterized by its violet belly. As its name suggests, it has iridescent violet fans on the sides of the head. The rest of the plumage is a rather uniform glittering green.

WHITE-BELLIED WOODSTAR *Chaetocercus mulsant* 7.7cm

Male   Female

One of six species of woodstar in Peru, this tiny bird flies with wing-beats so fast it has a bee-like appearance. In typical woodstar fashion, the male is very different from the female, having a deep purple-pink throat and deep green iridescent upperparts. The female has white underparts with a banded pattern, as shown in the photograph. She also has deep green upperparts. A common little hummingbird of the upper cloud forest edge on the eastern Andean slope between about 2000m and 3000m, where there is a profusion of flowering shrubs. Commonly seen on the upper Manu road.

GIANT HUMMINGBIRD *Patagona gigas* 18.5cm

This, the largest of all the hummingbirds, has little vibrant colour in its plumage. It has a bright whitish rump but is otherwise generally a brownish bird, best characterized by its size and relatively slow wing-beats. It can be seen in gardens in and around the city of Cuzco, where it feeds largely on the flowers of Prickly Pear. Will defend its territory with gusto, as will many hummingbird species, although, given the size of this one, it is largely successful in seeing off other birds and easily chasing them away from its precious food sources. Unmistakable, due mostly to its size. Quite a common bird in the highlands throughout the country, from about 1000m up to 4500m.

BLUE-CHINNED SAPPHIRE *Chlorostilbon notatus* 8cm

This spectacularly shiny little hummer is a bird of the Amazonian lowlands. It is fairly common, usually being seen at the forest edge and sometimes quite low down. The male is mainly bright glittering green, the throat being somewhat bluer, although this is usually rather hard to discern. It has a quite short, straight bill with a red base to the lower mandible and a rounded blue-black tail. The female has dull white underparts. Rather similar to several other species, especially the Blue-tailed Emerald *Chlorostilbon mellisugus*, but that species is smaller, has a forked tail, and the female has a whitish stripe behind the eye, as well as being greyer below.

COLLARED INCA *Coeligena torquata* 11cm

This lovely and rather flashy hummingbird lives in cloud forest habitats on the eastern Andean slope between roughly 2000m and 3000m. The bird illustrated is the more northern of two species of inca in Peru. The inca from the south has a pale orange collar and has recently been reclassified as a separate species, which is now known as Gould's Inca *Coeligena inca*. This species is typified by the white lower throat and breast, contrasting sharply with the deep, almost blackish-green body plumage, and further set off by the white flashes in the tail. It has a rather long and very straight bill. The female is rather similar to the male, but is a little duller.

TYRIAN METALTAIL *Metallura tyrianthina* 10cm

One of the commonest hummingbirds in the highlands of Peru. Inhabits cloud forest and in particular forest-edge habitats, as well as more open scrubby areas, wherever there are flowering trees and shrubs. The males are variable in appearance as there are several subspecies in different parts of the country. They are generally only a moderately iridescent green, but with a brighter glittering throat. The tail varies in colour, being deep blue in the south, violet blue further north, and in the far north a rich copper colour. The bill is short and straight. The female, shown in the right-hand photo, is a pale buffish colour below with a white spot behind the eye.

AMAZILIA HUMMINGBIRD *Amazilia amazilia* 9cm

This lowland Pacific coast hummingbird is quite common in the gardens of Lima and likely to be the first of the 127 species of Peruvian hummingbirds to be encountered on arrival in the country, as it is found around many Lima hotels, where it feeds on ornamental plants. This and the Oasis Hummingbird *Rhodopis vesper* are the two hummers most likely to be seen from the swimming pool. The latter, however, has a completely different appearance, with a rather long, decurved bill, purple throat and white belly. The Amazilia is unmistakable, as no other Peruvian hummer resembles it. It can be confusing, however, as it has several forms: some with a green chest, others white, and even blue in the south.

MASKED TROGON *Trogon personatus* 26cm

The only truly highland trogon. Very similarly patterned to some other members of the family, with a sharply demarcated white band across the breast. With his striking dark green and red plumage, the male is unmistakable in this bird's cloud forest home. The hooting call is both unique and distinctive in the higher levels of its altitudinal range. Occurs in the cloud forests of the eastern Andean slope from about 700m almost up to the treeline. The very similar Collared Trogon *Trogon collaris* occurs sympatrically at lower elevations. The females of both species are very different from the males, having rich brown upperparts, and the female Masked having a blacker face and throat than the female Collared.

(AMAZONIAN) WHITE-TAILED TROGON *Trogon viridis* 23cm

One of the yellow-bellied group of Peruvian trogons, which includes the Violaceous Trogon *Trogon violaceus*, which is very similar but has a yellow eye-ring. Both of these beautiful birds are quite common in the lowlands of the east. Characterized by the white undertail of the male, although the tail of the female is barred on the underside. Both sexes have a pale blue eye-ring. As with other members of its family, it has a far-carrying and distinctive call, which is easily imitated and will draw a bird almost automatically to the observer. The upperparts of both species are beautiful iridescent shades of deep purple and blue.

RINGED KINGFISHER *Ceryle torquata* 38cm

The largest of all the Neotropical kingfishers, common along lowland rivers and other wetlands east of the Andes. Easily identified by its large size and coloration, and very large bill. Both sexes have grey upperparts. The male has a completely reddish belly, and whilst the female has similar rufous underparts, it also has a grey band across the breast, separated from the belly by a narrow white band. This bird's size and distinctive pattern make it hard to miss. It also has a loud rattling call, which usually announces its presence in advance. A spectacular member of the riverside bird community.

AMAZON KINGFISHER *Chloroceryle amazona* 27cm

By far the largest of the three Amazonian green kingfishers, with a massive dagger-like bill. Both sexes have beautiful, deep green upperparts and largely white underparts, and in common with the Green Kingfisher *Chloroceryle americana*, both males are typical of this group of kingfishers in having chestnut-orange breast bands. Has a fast and powerful flight, although not as speedy as its smaller relative, the Green Kingfisher. It differs from the rather smaller Green-and-rufous Kingfisher *Chloroceryle inda* in that the latter species has wholly rufous underparts. The female Amazon has green sides to the breast, whereas the Green has a complete green breast band. A common bird of the rivers of the eastern lowlands and therefore one of the species regularly seen from boats.

GREEN KINGFISHER *Chloroceryle americana* 20cm

Rather similar to the larger Amazon Kingfisher *Choroceryle amazona*, and also quite common but generally less conspicuous, partly owing to its smaller size. Tends to live along smaller streams than its larger congener, and often seen flashing past in very rapid flight. When seen at close range its beautiful, deep green iridescent plumage becomes apparent. This species also has a shorter crest than its larger cousin. It possibly has the largest bill of any of the Neotropical kingfishers relative to its size. Like most other kingfishers, it nests in a horizontal tunnel excavated into a river bank.

AMERICAN PYGMY KINGFISHER *Chloroceryle aenea* 13.5cm

The smallest of the American kingfishers and perhaps the most attractive. Can be remarkably tame when encountered in a small boat. Often sits quietly by the side of the smallest of streams and under overhanging vegetation, which makes it very difficult to spot despite its brilliant multicoloured plumage, unless seen at close range. It can also be hard to see in flight, being so small and flying very rapidly. Its deep green head and upperparts, separated by an orange-cream collar, and deep reddish-chestnut underparts, make for a very striking pattern on this tiny kingfisher. Takes very small fish, as expected from a kingfisher of its size.

BLUE-CROWNED MOTMOT *Momotus momota* 41cm

This member of a very attractive family is probably the commonest motmot of the region, and is found in a variety of wooded habitats, including secondary forest. The beautiful electric blue stripes on the head and the black mask are the most obvious features of this lovely bird. It has warm, chestnut-hued underparts, contrasting with vivid deep green upperparts. The motmot's tail is characterized by the bare quills with spatula-shaped tips. These are actually made by the bird itself, by plucking out the feather barbs to leave a bare shaft. This is in common with some other members of the family, as is also its habit of sitting quietly and waiting for prey to show.

BROAD-BILLED MOTMOT *Electron platyrhynchum* 35cm

This is an unusual motmot in having a completely different call from the typical '*mo mo*', which gives the motmots their name. Instead, its call is a rather strange, growling sound that sounds a bit like a cicada, or even a short burst of a distant chainsaw. Its orange-rufous breast with a black spot in the centre is also distinctive, but not entirely diagnostic, as it closely resembles that of the Rufous Motmot. That species also has a thicker and less down-curved bill, more extensive rufous on the underparts and a smaller black breast spot. It is also a considerably larger bird.

HIGHLAND MOTMOT *Momotus aequatorialis* 48cm

As its name suggests, this species (once 'lumped' with the Blue-crowned Motmot) is a bird of the highland cloud forests of the eastern slope of the Andes, between about 1500m and 3000m. In common with many other birds of the higher elevations, it is larger than its lowland congener. It has a pale, electric blue forecrown, stripes along the sides of its head, and greener underparts than its smaller lowland cousin. Its red eyes add to the extravagant head pattern. Has a tendency to sit in trees or on rocks beside streams, where it can be very obvious. Also perches at the edges of cloud forest and is most often seen along montane roads.

PARADISE JACAMAR *Galbula dea* 31cm

One of the most elegant of the jacamars, with a very beautiful shape. Its long and very pointed bill and elongated tail give it a very distinctive profile. Its plumage is largely black but with a striking and contrasting white throat. It is a bird of the lowland rainforest, where it sits high in the canopy waiting and watching for prey. It is rather inconspicuous at times, with little obvious colour, and it perches high up. In the eastern and central Amazon it is locally quite common, but in Peru it is a rather difficult bird to find, being much more localized. Uncommon to rare in most parts, although it is quite widely distributed throughout the lowland forests.

PURUS JACAMAR *Galbalcyrhynchus purusianus* 20cm

This strange-looking bird with its huge pink bill is a specialist that lives only in areas of dead trees near rivers in lowland rainforest. Its preferred habitat is formed by the flooding of these areas during the rainy season. Can be seen in such conditions along the Rio Madre de Dios in Cuzco province, particularly at Blanquillo, just downstream from Boca Manu. Has a loud and distinctive trilling call, which often alerts one to its presence. Very similar to the White-eared Jacamar *Galbalcyrhynchus leucotis*, which occurs only in the far north of the country but lacks the white 'ears' (actually white patches on the sides of the head behind the eye). The two species are clearly very closely related.

WHITE-NECKED PUFFBIRD *Notharchus macrorhynchos* 25cm

This large and strikingly patterned puffbird is a spectacular creature, and like many other members of the family, it is a 'wait and see' hunter. It sits quietly in the mid-canopy, waiting to ambush prey, and is capable of taking many larger prey items than are tackled by other puffbirds, including large insects, lizards, frogs, and even small snakes. It has a white throat and forehead, a thick black band across the breast, finely barred flanks and black upperparts. Although widely distributed in the eastern lowland forests, it is a rather uncommon bird throughout most of its range.

WHITE-EARED PUFFBIRD *Nystalus chacuru* 21.5cm

This is an open country species which is much more common further south, in Bolivia. It occurs in the Pampas del Heath in the extreme southeast of Peru and also, rather uncommonly, in the drier and more open areas of some of the intermontane valleys on the east side of the Andes. Perches up in prominent positions, such as fence-posts, as shown in the photograph, where it is obvious and thus easily seen. As is usual with the puffbirds, it often sits for long periods waiting for an unfortunate item of prey to come along. This particular species usually pounces to the ground from its vantage point in the manner of a shrike, of which there are no examples in Peru.

BLACK-STREAKED PUFFBIRD *Malacoptila fulvogularis* 23cm

This lovely but rather subtly plumaged puffbird is unusual in that it occurs at higher altitudes than most of its relatives. It is found in lower temperate or upper tropical level cloud forest at elevations between about 1500m and 2100m. Although rather closely resembling several other puffbird species, it does not overlap with them in its altitudinal range. The orange-buff throat with white moustachial marks is diagnostic. It has a dark, streaked head, brown upperparts and a boldly streaked belly. An uncommon bird, but perhaps seen less often than it might be simply because it sits still for long periods, so does not attract attention.

COLLARED PUFFBIRD *Bucco capensis* 18.5cm

This beautiful puffbird is one of the most striking members of the family, with its bright orange-rufous head and large, bright red bill and black breast band. A rather rare bird in Peru, it occurs only in the northeastern Amazonian region, near Iquitos. However, as with other members of the family, it sits motionless for long periods and is probably very often overlooked, preferring to lurk in mid-storey cover, such as vine tangles.

BLACK-FRONTED NUNBIRD *Monasa nigrifrons* 28.5cm

Although belonging to the puffbird family, the members of the genus *Monasa* are very different, having a characteristic uniform blackish plumage. There are three species in Peru, of which this is the most common. All three are inhabitants of the eastern forested lowlands and the lower foothills. This species is identifiable by its red bill and charcoal-black plumage with deeper black face. The White-fronted Nunbird *Monasa morphoeus* also has a red bill and also a white face, as one would expect from its name. Likewise, the Yellow-billed Nunbird *Monasa flavirostris* has a very distinctive yellow bill, as well as white on the scapulars.

SWALLOW-WING *Chelidoptera tenebrosa* 16cm

This is a rather aberrant member of the puffbird clan and one that displays rather different behaviour from all the other members of the family. Its *modus operandi* is to sally out from high perches to capture flying insect prey. In this respect its behaviour more closely resembles that of a large flycatcher, such as a kingbird or a kiskadee. It has a very short tail and very broad, blunt wings, which it uses to glide back to its vantage point after a foray. Deep blue-black above with a whitish vent, it may appear uniformly black when silhouetted against the sky. Most often seen from boats on the eastern lowland rivers.

GREY-BREASTED MOUNTAIN-TOUCAN *Andigena hypoglauca*
46cm

This beauty is one of four mountain-toucan species, of which three occur in Peru. Its distinctive yelping calls are easily imitated, and this bird will often come to investigate if this is done. It has a very attractive plumage pattern with a contrasting array of colours: a blue-grey body, blackish head, and olive-green upperparts, with a bright yellow rump. It also has, as the photograph shows, an amazingly ornate bill, with a combination of black, yellow and red. It can be seen along the Manu road near Cuzco at about 2600m elevation, in the cloud forest above the point known as Pillahuata.

TOCO TOUCAN *Ramphastos toco* 55cm

This species, although not the largest of the toucans, is certainly the one with the largest bill, and is a bird of the more open country of the eastern lowlands. Generally rare in forested areas but more regularly seen in the deforested southeast corner of the country. This species is an opportunist and largely eats fruit, but will also predate just about anything it can catch. I have personally seen it smash its way into a thornbird's fortified nest and eat all the chicks, using the huge yellow, black-tipped bill that is a diagnostic feature of this species. At close range the vibrant violet-purple bare skin around the bird's eye can be noticed.

CHESTNUT-EARED ARACARI *Pteroglossus castanotis* 43cm

The aracaris, although clearly members of the toucan family, are very different from the larger, mostly black-and-white *Ramphastos* species, having contrasting yellow and red-banded underparts, and mostly blackish upperparts. They are also typified by having a long tail, particularly conspicuous in flight, which is rather laboured. They are often seen crossing rivers in small, loose flocks, usually one at a time. This species is usually the commonest member of the genus and the one most often seen by visitors to the lowland forests of the east. Seven members of the aracari tribe have been recorded in Peru.

CREAM-COLOURED WOODPECKER *Celeus flavus* 26cm

This strange-looking member of the woodpecker family is a denizen of the lowland forests of the east. The *Celeus* genus is known for the rather flamboyant, even eccentric appearance of many of its members, but this species is, literally, out on a limb in this respect. Few woodpeckers even remotely resemble this bird, with its odd yellow-cream body plumage, contrasting dark wings and shaggy triangular crest. However, the male has the typical red moustache of the Celeus group (female shown in photograph). Is very fond of raiding arboreal ant and termite nests to feed on the larvae, and has a loud and high-pitched ringing call. It is an unmistakable bird.

CHESTNUT WOODPECKER *Celeus elegans* 27cm

This gorgeous, richly-coloured woodpecker is a typically ornate member of the *Celeus* genus and locally a fairly common bird of the terra firme forests of the eastern lowlands. Owing to the nature of its habitat, it is sometimes difficult to see but will come to feeders that offer fruit. The male has a bright red malar stripe, although this is absent in the female, as is usual among the *Celeus* genus. Its plumage is mostly rich chestnut with a contrasting yellow-buff rump. Very similar to the Scaly-breasted Woodpecker *Celeus grammicus* but, as its English name indicates, that species has black scaling on the throat and breast and finer barring on the upperparts.

RED-NECKED WOODPECKER *Campephilus rubricollis* 34cm

This large and boldly marked woodpecker is a fairly common inhabitant of the lowland Amazonian rainforests. It has a distinctive crested red head and neck, dark upperparts and chestnut underparts. The male has a small black-and-white spot on the ear coverts and the female (shown in photograph) a white stripe from the base of the bill. This is rather typical of the *Campephilus* woodpecker group. Its presence is often indicated by the distinctive sound of its 'drum', which is typically made by banging very loudly and rapidly twice on a tree, and could be transcribed as '*ba bang*'. this sound can be heard more than a kilometre away in still conditions. Its call is a loud rasping '*kyaah*'.

YELLOW-TUFTED WOODPECKER *Melanerpes cruentatus* 20cm

As is usual with the *Melanerpes* genus, this species roams around in small but noisy flocks and is usually heard first, since it makes a very distinctive high-pitched 'twittering' call. It also has the usual, rather clown-like, plumage so typical of this group of wood- peckers, being mostly blue-black with a red crown and a yellow eye-ring, and a stripe running back from the eye. The rump is contrastingly white and the belly red, with fine black-and-white barring on the flanks. This species is an inhabitant of the lowland rainforests of the east, and is a common species in suitable habitat. It will readily come to bird feeders if tempted by fruit.

WHITE WOODPECKER *Melanerpes candidus* 20cm

This woodpecker has a totally different lifestyle from all other species of woodpecker in Peru. Restricted to the southeastern corner, its preferred habitat is open country, especially in the Pampas del Heath and cleared areas south of Puerto Maldonado. Here it often utilizes fence posts and dead trees. Typical of the *Melanerpes* woodpeckers, it roves around in flocks of between about five and ten birds, covering large distances in search of foraging opportunities. With its striking plumage, combining a white head and underparts with blackish back and wings, it is an unmistak- able and highly attractive species of bird.

GOLDEN-OLIVE WOODPECKER *Piculus rubiginosus* 23cm

This woodpecker in some ways resembles the Black-necked Woodpecker *Colaptes atricollis*. However, it is a bird of very different habits, as it prefers to live in woodland rather than the arid cactus-strewn slopes favoured by the Black-necked Woodpecker. That species has a markedly barred back, but has a very similar and striking head pattern, with a dark throat and forehead and white cheeks, which is largely what gives it the similar appearance. Although it occurs on both sides of the Andes and is regularly seen in the forests of the lower eastern Andean slope, it is probably more common in the forests of the dryer Pacific slopes of the north.

MONTANE WOODCREEPER *Lepidocolaptes lacrymiger* 20cm

One of only two woodcreepers that occur in the temperate zone of the eastern Andean slope between about 1500m and 3000m, this is by far the commoner of the two species. The other species is the Tyrannine Woodcreeper *Dendrocincla tyrannina*, which occurs at similar elevations but is much rarer. It is also larger, with a heavy, dark bill and looks very different, being plain brown with rustier wings and tail. This species also has rufous wings and tail, but is very obviously streaked below, has a spotted crown and a plain mantle, a slender decurved bill and is a considerably smaller bird.

PLAIN-BROWN WOODCREEPER *Dendrocincla fuliginosa* 20cm

This species is fairly common in the forests of the eastern lowlands and lower Andean foothills. As its English name suggests, it is a bird with little in the way of distinguishing markings in its plumage. It has the typical rufescent wings and tail of a woodcreeper, and is slightly paler and rather warmer-hued on the underparts. It displays a rather subtle face pattern, having somewhat greyer ear-coverts and a darker but not particularly significant malar stripe. Sometimes attends swarms of army ants, but is not as closely tied to this habit as its close relative, the White-chinned Woodcreeper *Dendrocincla merula*, which is very similar but has pale blue eyes and, of course, a white chin.

COASTAL MINER *Geositta peruviana* 16.5cm

This is one of seven species in a group of small terrestrial birds, which are members of the ovenbird family, the Furnariidae. They are typified by living on open ground, which is usually dry and sandy or rocky. They are also characterized by generally having earth-coloured plumage to match their environment and, of course, provide camouflage protection. This species, as its name suggests, occurs in the arid coastal zone, andusually lives in places where there is no vegetation whatsoever. It is a common bird in this habitat and can often be seen by the side of the Pan American highway. It is also a common sight around archaeological sites, such as Chan Chan and the nearby Moche pyramids.

BAR-WINGED CINCLODES *Cinclodes fuscus* 18cm

This is the commonest and most widespread member of the *Cinclodes* genus, members of the ovenbird family, the Furnariidae. There are three races in Peru. It is regularly seen along the sides of roads in the highlands, from about 2500m up to about 5000m. Although it occurs in a variety of open habitats at high elevations, it is especially fond of damp areas such as marshes, streams, lake margins and roadside ditches. At first glance it appears to be a rather plain, brown bird with whitish underparts, but is characterized by the distinctive wing pattern. It has a prominent wing-bar, which is very obvious when seen in flight. Depending on the race, this bar can be any shade between buff and white, when it closely resembles the White-winged Cinclodes *Cinclodes atacamensis*.

WHITE-BELLIED CINCLODES *Cinclodes palliatus* 23cm

This large species of cinclodes is a very rare and local bird, being known from only a few localities in central Peru at high altitude, and in quite specific habitat. It is an inhabitant of *Distichia* bogs and adjacent areas, between about 4400m and 5000m. The total population may be only a few dozen pairs. It is a striking bird, being a rich, deep chestnut colour above with a paler greyer head and contrasting white underparts. Like other cinclodes, it is a ground-dwelling bird, feeding in the bogs and singing from rocky outcrops overlooking them. Most easily seen only a few hours' drive out of Lima, along the road to Laguna Marcapomococha, in the vicinity of Milloc bog.

YELLOW-CHINNED SPINETAIL *Certhiaxis cinnamomea* 14cm

Occurring in the tropical lowlands east of the Andes, this member of the family of furnarids is always seen close to water, and in waterside or floating vegetation along rivers or on oxbow lakes. As well as being a characteristic bird of its habitat, it is typified by its yellow upper throat, although this feature is generally difficult to discern. Otherwise it is very similar to the Red-and-white Spinetail *Certhiaxis mustelina*, but that species lacks the yellow throat and has blacker lores and also a reddish crown, which is the same colour as the mantle. In this species the forecrown is grey. Although fairly common in its habitat, it is sometimes rather unobtrusive in the thick waterside cover.

WARBLING ANTBIRD *Hypocnemis cantator* 12cm

Probably one of most absurdly named birds of Peru, as the song of this species certainly cannot be described as warbling. In fact, it has a very distinctive voice, which is a series of rasping notes, descending in scale. It is a common bird of the eastern lowlands, where it inhabits all manner of wooded habitats, but is commoner in successional growth, marginal areas along rivers, secondary growth and forest edges. It thrives in such areas, which are largely shunned by many other antbird species. There are four subspecies in Peru, and the races *subflava* and *collinsi*, from the southeast, have a yellowish belly and may in fact represent a separate species.

AMAZONIAN STREAKED-ANTWREN *Myrmotherula multostriata* 9.5cm

A rather uncommon antwren of the humid forests of the eastern lowland, but locally more common near water. Almost always only seen in vegetation beside streams or oxbow lakes, where its presence is usually only apparent when its distinctive lilting song is heard. This could be transcribed as *'pur tsi pur tsi pur tsi pur tsi'*, wavering up and down the scale. Most often and best seen from a small boat. The male is striped black and white, but the female has an attractive orangey-rufous head, finely streaked with black.

RUSTY-BACKED ANTWREN *Formicivora rufa* 12.5cm

This lovely and striking antwren is unique in Peru in having this particular plumage pattern. The male has a black throat and breast, which narrows as it runs down to the belly, a white border separating this from the rufous upperparts. The undertail coverts and flanks are a pale yellowish-buff colour. The tail is tipped with white and the wings are adorned with white spots and wing bars. The female is very different from the male, having streaked underparts. A fairly common bird of the more open scrubby habitats on the lower foothills on the eastern slope, and especially the savannahs of the lowland, such as Pampas del Heath.

TAPACULO sp. *Scytalopus sp.* 12cm

This is just one of the *Scytalopus* tapaculos, which have evolved into many very similar species; their taxonomy is in constant flux. They rarely fly, preferring to run, and sing from only moderately elevated perches. Many species are restricted to small geographical areas and also to very narrow altitudinal ranges. Highly secretive, most are found in cloud forest between 1000m and 4000m altitude. Best identified by voice, usually a simple chipping or trilling song.

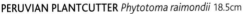

PERUVIAN PLANTCUTTER *Phytotoma raimondii* 18.5cm

This strange bird is a member of a family consisting of only three species in the world. This particular species, which is endemic to Peru, is in dire threat, as it lives only in areas of desert with patches of scrubby, woody plants. These are being rapidly exploited for firewood by the burgeoning human population. The species is seemingly related to the cotingas, but that large and diverse family is probably one of the most contentious groupings of birds, as it is made up of so many different forms. This particular group are characterized by their stubby bills, mixed-up pattern of greys and orange or reddish tones, and white wing-bars.

SPANGLED COTINGA *Cotinga cayana* 20cm

A particularly beautiful cotinga, which is most often seen from canopy observation towers, as it spends its entire time roaming through the rainforest treetops, from one fruiting tree to another. It is most easily seen in the early morning sunshine, when it perches out in the open and basks, before setting off to forage. As a fruit-eater it does not need to be up too early, because it knows most of the fruit will probably still be there and will be riper later. However, if there is an influx of migrant Eastern Kingbirds *Tyrannus tyrannus*, a regular food source can be stripped of fruit in a very short time.

PURPLE-THROATED FRUITCROW *Querula purpurata* 27cm

This large member of the cotinga family is fairly common in the lowland forests of the east. A fruit-eater, it wanders around high in the forest canopy in loose flocks, foraging from various trees. It is nearly always first encountered by its distinctive, soft but far-carrying mewing calls, which could perhaps best be transcribed as *'pew pew'*. Has a gentle, loping flight action. Its name is entirely descriptive as it is a quite crow-like bird, with an entirely blackish plumage, except for a lovely purple throat, although this can be hard to see in poor light. The female lacks this adornment, being more uniform.

ANDEAN COCK-OF-THE-ROCK *Rupicola peruviana* 30cm

One of the most iconic birds of Peru, and a species of the cloud forests of the eastern Andean slope. A member of the cotinga family, it is also known for its strange cat-like calls, which indicate the presence of a lek. Here the males gather to display, in the hope of attracting a mate. These rather noisy gatherings, with males bowing and wing-flapping to attract attention, are an unmissable specatacle. One very often visited location is on the road from Cuzco to Manu, just a few hundred metres from the aptly name Cock-of-the-rock Lodge.

WHITE-BEARDED MANAKIN *Manacus manacus* 11cm

This tiny member of the cotinga family is characteristic for its lekking behaviour, which involves a great deal of noisy wing-snapping and calling. The wings are used to make the very loud cracking noises that make the bird's leks so peculiar and obvious. At the same time, the males stick out their throat feathers to form something like a 'flag' or the 'beard', which gives the bird its English name. With a black back and cap, a white collar and underparts, the male is very dapper. The female is a rather anonymous olive-green bird, the most obvious feature being its bright pink legs, also a feature of the male.

GOLDEN-HEADED MANAKIN *Pipra erythrocephala* 9cm

A typically spectacular manakin, which occurs in the lowland forest east of the Andes, but is not found in the southeast. Its main feature is the male's brilliant, golden-yellow head, contrasting with the uniform black of the rest of its plumage. The male also has startling red and white thighs, which although exposed when displaying, are otherwise usually concealed. The female is, as usual with manikins, a drab creature compared to the male, and in this species a uniform olive colour with a paler belly and pale bill. The display at the lek is a noisy and flashy affair with males trilling and buzzing. Most commonly seen near fruiting trees.

CINEREOUS MOURNER *Laniocera hypopyrra* 20cm

A rather unusual member of the tyrant flycatcher family. At first glance, it appears to be just a rather anonymous grey bird, but with good views it can be seen to have two lines of cinnamon-coloured spots on the wings. These spotted wing-bars are formed by the tips to the lesser and median coverts. The tertials and the tail feathers are also tipped with the same contrasting colour. Generally a scarce and inconspicuous bird, it is found in the lowlands of the east in the middle storey of *terra firme* forest. Young birds are confusing, with a rufous and black spotted pattern on the breast. The song is a repeated series of high-pitched ringing calls.

LULU'S PYGMY TYRANT *Poecilotriccus luluae* 10cm

This delightful little bird is endemic to the country. Only recently described, it is a bird of secondary growth and bamboo rather than true forest habitats. It is rather rare, but probably commoner than is currently understood, owing to the remote nature of its mountainous range. Occurs between about 1800m and 2200m altitude, and is probably best looked for in patches of bamboo near the pass at Abra Patricia, in north-central Peru. Like most other members of the genus (of which there are four in Peru), it has a rather intricate plumage pattern. This is generally typified by yellow underparts and an ornate head pattern, the exception being the Black-and-white Tody-Tyrant *Poecilotriccus capitalis*, which is, of course, black and white.

CLIFF FLYCATCHER *Hirundinea ferruginea* 18.5cm

This quite large flycatcher specializes in living on rocky roadcuts and cliffs, as its English name suggests. It is a slim and rather streamlined bird, with richly coloured underparts and a grey head marbled with white. Sallies out from its rock perch to capture prey, looking rather like an old world bee-eater with its bright rufous underwings. Also has a distinctive gliding flight, which adds to this appearance. Quite common on the eastern slopes of the Andes at middle altitudes and can, for example, be seen readily along the Manu road out of Cuzco in the lower temperate and sub-tropical zones, especially where there are rocky outcrops.

VERMILION FLYCATCHER *Pyrocephalus rubinus* 15.5cm

A common bird of the arid Pacific slope up to about 2000m, and also a migrant to the eastern lowlands during the austral winter. Can commonly be seen in and around Lima. The male is normally a bright but deep red below with a contrasting dark upper side. This is variable in tone, but can be almost completely blackish in the race *obscurus*, which occurs from Lima south to the Chilean border. The female is pale below with varying amounts of pink and sometimes yellow on the lower belly and vent, and usually some streaking on the breast. The female of *obscurus* can appear completely dark, only displaying some pink on the vent.

BRAN-COLOURED FLYCATCHER *Myiophobus fasciatus* 12.5cm

A common flycatcher on both sides of the Andes. The three forms are currently treated as one single species, but may be split in future. The form from the eastern Andean foothills is also a migrant from the south in the austral winter. The northern form is a dull olive-brown. The eastern Andean populations are a dull rufous-brown and are typified by the obvious wing-bars and in particular the streaking on the breast, which is often quite bold and is a diagnostic feature of this race. Birds on the Pacific slope are brighter, with pale cinnamon underparts, rufous wing-bars and lacking the streaking.

BROWN-BACKED CHAT-TYRANT *Ochthoeca fumicolor* 15cm

One of a total of twelve species of chat-tyrants in Peru. The genus is generally typified by double wing-bars and supercilia. A bird of the cloud forest edge, treeline scrub and other wooded habitats between about 2600m and 4200m, and quite common in its preferred scrubby habitat within its altitudinal range. The race *brunneifrons*, in the north of the country, has a long buff supercilium, and in the southern part of Peru the race *berlepschi* has a narrow whitish supercilium. Both subspecies have rusty wing-bars, but these are rather bolder in the northern race. Typically rather tame.

SOUTHERN BEARDLESS-TYRANNULET *Camptostoma obsoletum* 10cm

One of the most commonly seen of the tyrannulets, as it does not require any specialized habitat, living anywhere with trees, including urban gardens, parks and agricultural areas. Since it is not a forest bird, it is uncommon in the eastern lowlands. However, it is a common bird on the Pacific coastal plain, and also in semi-arid intermontane valleys. There are several different subspecies, which differ in appearance but all share a characteristic little pointed crest, obvious wing-bars and cocked tail. Usually a dull brown colour above, some greyer than others and some more olive-hued.

BLACK PHOEBE *Contopus nigricans* 12.5cm

This flycatcher is a common bird along the rivers and streams of the eastern Andean foothills between about 500m and 2500m, where it can be quite obvious. This is due to its habit of perching out in the open, often on rocks in or by the side of a stream. It is also seen around buildings, and commonly perches on and nests under bridges or culverts, where it is easily seen by anyone passing by. It typically flycatches from these low perches, and is quite tame and rather endearing. An attractive bird, with its dapper black body plumage and white wing-bars. The fairly long tail is frequently flicked up.

LITTLE GROUND-TYRANT *Muscisaxicola fluviatilis* 13.5cm

A bird of the riversides and river sandbanks of the eastern Amazonian lowlands; all the other ground-tyrants normally occur in the highlands or on the Pacific slope of the Andes. This species usually only ranges up to about 1000m at most, rarely a little higher, where it would usually be seen in open areas near water or on unpaved roads. A rather plain bird, this small and terrestrial flycatcher is perhaps most easily recognized by its nondescript appearance, as it runs around on the ground. It resembles the Spot-billed Ground-Tyrant *Muscisaxicola maculirostris* in size and general appearance, but that species is a bird of higher altitudes and unlikely to overlap with the range of this bird.

DARK-FACED GROUND-TYRANT *Muscisaxicola macloviana*
16cm

One of twelve species of this genus of terrestrial flycatchers in Peru, this species visits in the austral winter from its breeding grounds in the far south of Patagonia and the Falklands. Commonly seen between April and September along the coast and foothills, in places such as Lomas de Lachay. As with other ground-tyrants, it is a rather nondescript and uniform bird, and has a similar upright profile when on the ground. It can be distinguished from other ground-tyrants by its dark brown face. Sometimes seen in roving flocks, searching for suitable areas on which to forage. Roams as far north as Trujillo, northern Peru, hence a true long-distance migrant.

ANDEAN NEGRITO *Lessonia oreas* 12.5cm

A rather unusual, small and stocky terrestrial member of the tyrant flycatcher family. A very active little bird. The male is highly distinctive, with its black head and underparts and rich chestnut mantle; the female is a rather duller and browner version of the male. Occurs at high altitudes, typically between 3000m and 4000m, usually in open, damp areas in the southern half of the country. It is most commonly seen along the margins of highland lakes and streams, and always on the ground. As with a number of high Andean birds, it has been recorded rarely at sea level. Easily seen at Huacarpay lakes near Cuzco.

WHITE-HEADED MARSH-TYRANT *Arundinicola leucocephala*
12.5cm

This strangely plumaged and very distinctive little bird is found in waterside habitats in the lowlands of the northeast. It is rather uncommon, and most often seen in damp grassy places and low scrubby areas near lakes or rivers. It catches its prey by sallying from low perches, such as vertical grass or reed stems, or small trees and bushes. The male is absolutely distinct from all other birds, with its black body plumage and white head. The female is less distinctive, being a dull grey-brown above and whitish below with pale sides to the head and a white forehead.

WHITE-TAILED SHRIKE-TYRANT *Agriornis andicola* 28cm

A very large member of the tyrant flycatcher family, with a heavy hooked bill and a pale base to the lower mandible. The bill of the very similar but smaller and much commoner Black-billed Shrike-Tyrant *Agriornis montana* lacks the pale base, but both species share the general characteristic of the obvious white tail and dark central feathers. The throat of this species is more heavily streaked, however. It is a very rare and local bird in the northern highlands, and in recent years has suffered a serious decline in its population. The reasons for this remain unclear. Can at present (2006) be seen along the road from Celendin to Cajamarca in northern Peru.

RUFOUS FLYCATCHER *Myiarchus semirufus* 18cm

The rarest *Myiarchus* flycatcher in Peru. Not only does it have the most restricted range but it also requires a desert habitat, where it is to be found in scattered patches of scrubby trees. These are under great threat from felling for firewood, a problem that is also affecting the Peruvian Plantcutter *Phytotoma raimondii*, a species that occurs alongisde this bird in the same areas. Very different in appearance from all other *Myiarchus* flycatchers, this is a deep, rich rufous-coloured bird, and unlike the mainly plain-brown colouration of most of this family. It is unique in its environment, and can be seen particularly near the town of Rafan.

DUSKY-CAPPED FLYCATCHER *Myiarchus tuberculifer* 17cm

Just one of several members of the very confusing genus *Myiarchus*, rather large members of the family of tyrant flycatchers, the *Tyrannidae*. This species, however, is not as similar to other members of the genus as some are. It can usually be separated by its very dark crown, which is normally quite apparent and can sometimes be almost black. It contrasts with the paler olive-brown mantle. The flight feathers are indistinctly edged with rufous. The species occurs on both slopes of the Andes, but at much higher elevations in shrubbery on the Pacific slope. On the eastern Andean slope it occurs in forest-edge and semi-wooded habitats up to about 1800m.

BROWN-CRESTED FLYCATCHER *Myiarchus tyrannulus* 19.5cm

One of the largest members of the genus and the only species regularly seen in Peru with very obvious rufous edges to its primaries. Although the Dusky-capped Flycatcher *Myiarchus tuberculifer* shows some rufous in the wing, the Brown-crested never displays the contrasting dark head of that species and is bigger. Locally fairly common in parts of the north and east, notably in the upper Marañon valley up to about 1700m, and apparently occurs in southeast Peru as an austral migrant. Very similar to the even larger Great Crested Flycatcher *Myiarchus crinitus* from North America, which has been recorded from the lowlands in the north. That species shows more rufous edgings in the tail and has a different call.

GREAT KISKADEE *Pitangus sulphuratus* 22cm

One of the largest and most conspicuous of the yellow-breasted tyrant flycatchers. The loud calls lend the species its name, 'kis-ka-dee', one of the most characteristic and familiar sounds in many habitats all over the eastern lowlands, up to a height of about 1000m. Occurs commonly in towns and villages, all manner of open areas and marginal habitats, and particularly likes areas near water. Very easily recognized. Apart from the bright yellow belly, its most obvious feature is the vibrant black-and-white head pattern. It has a heavy black bill and rufous margins on the wings and tail. It also has a semi-concealed yellow crown-stripe, which can flare in threat or display, as shown in the photograph.

CROWNED SLATY FLYCATCHER *Griseotyrannus aurantioatrocristatus* 18cm

This species is a rather anonymous member of the tyrant flycatcher family, being a rainforest canopy bird and displaying no bright colours. It migrates long distances, from southern South America to the lowland rainforests of the east and further north. This bird was photographed when it landed on a ship in the middle of the Amazon whilst migrating. Its English name indicates that it has a distinctive head pattern; this is a semi-concealed yellow coronal-stripe bordered with black, a feature rarely seen in the field as the bird usually lives in the top of the highest forest trees. Most regularly seen from purpose-built canopy viewing platforms, such as the one at Manu Wildlife Centre on the upper Rio Madre de Dios.

TROPICAL KINGBIRD *Tyrannus melancholicus* 20cm

One of the commonest and most regularly encountered birds of Peru, this is a bird whose identity should be quickly learnt as it is a ubiquitous species in the eastern lowlands and also in the west as far south as Lima and beyond. Once learnt, it is easily recognized. Seen in both city and countryside, it is generally referred to by the abbreviation 'T.K.' and has a distinctive trilling call. Its profile when perched is upright and streamlined, and it often sits on telegraph wires from which it sallies after its prey. Regularly chases other, much larger birds, including raptors; this behaviour has led to the derivation of the name kingbird. Rather plain grey on the head with similarly plain upperparts and pale lemon yellow below.

FORK-TAILED FLYCATCHER *Tyrannus savana* 20–40cm

The wide disparity in the figures given for the length of the bird indicates the length of the tail in the adult male. This striking flycatcher is a rather uncommon migrant to Peru, occurring in much larger numbers further east, where it is a very visible seasonal visitor. It breeds in southern South America and winters as far north as Mexico. Most likely to be seen flying over the lowlands east of the Andes, heading north in March or April and then south again in October or November. Easily recognized by its white underparts and especially the long, black forked tail. At rest, the pattern of the upperparts can be fully appreciated. The mantle is a soft dove-grey and the crown, nape and ear coverts black.

GREY-BREASTED MARTIN *Progne chalybea* 19cm

A large and common martin in most parts of its Peruvian range, which is mainly the eastern lowlands but also the northwest. Loud, conspicuous and inquisitive, it is one of the birds most obvious around towns and villages along the rivers of the Amazon lowlands. Often flies out from riverside towns to investigate and perch on ships plying the rivers of the region, usually calling excitedly – a rich, churring or chuckling sound. The male is deep blue above and white below with a grey breast. The female is brown above and pale below with a rather darker breast and throat. Brown-chested Martin *Progne tapera* is confusable, but has a pale throat and is always brown above.

WHITE-WINGED SWALLOW *Tachycineta albiventer* 13.5cm

This beautiful swallow is closely associated with the rivers and oxbow lakes of the Amazon basin, where it is a common and conspicuous bird. Many riverside dwellings have installed purpose-built bird houses for this species and it readily takes advantage of these nesting opportunities. Most often observed foraging over water, often flying very close to the surface. When seen perched, it is almost always on the branches of submerged trees or fishermens' marker posts protruding above the water surface. The turquoise-green upperparts, gleaming white underside and white fringes to the inner feathers on the upper wing are diagnostic. It is usually very confiding.

SOUTHERN ROUGHWING *Stelgidopteryx ruficollis* 13cm

One of the most commonly encountered swallows in the lowlands of the east and also along the north coast, south to about Cajamarca. Not a colourful bird, but quite distinctive in its sleek shape. It has long wings and a style of flight that is strong, fast and elegant. It is generally a rather dull-brown colour, with a brighter, paler rump and underparts and a warmer, rather cinnamon-buff throat. Often seen over water and commonly nests in riverbanks. It also burrows into road cuts, as with many other swallow species, making good use of man's presence. Its loud *'jreeep'* call is a familiar sound along the banks of the lowland waterways.

CORRENDERA PIPIT *Anthus correndera* 15cm

One of five species of pipit in Peru, and one of the more striking of a group which are notoriously difficult to identify. It has very well-defined blackish streaking on the breast and flanks. The mantle is generally buff, streaked with dark brown and displaying two long white stripes or 'braces' on the mantle. Quite a common bird in the puna grasslands, occurring almost as high as those grasslands exist. It forages exclusively on the ground, but has an aerial display flight. It shares its habitat with the Paramo Pipit *Anthus bogotensis* and Short-billed Pipit *Anthus furcatus*, both of which are less heavily marked and lack the dark streaking on the flanks.

WHITE-CAPPED DIPPER *Cinclus leucocephalus* 15.5cm

This is the only member of the dipper family in Peru and is unmistakable with its white breast and crown. Always associates with fast flowing mountain streams and rivers and will never be seen away from this very specific habitat. Most often seen perched on a rock midstream, or on a rock on the shore. This species does not actually dive into the water like most other dippers but always forages in or close to the rushing water. It is regularly seen from the train along the Urubamba River between Ollantaytambo and Aguas Calientes en route to Macchu Pichu. It flies close to the water surface with rapid wing beats and is often noticed only momentarily before it vanishes from view.

BLACK-CAPPED DONACOBIUS *Donacobius atricapillus* 22cm

A rather handsome bird that inhabits marshy habitats in the eastern lowlands of the country, in particular the wet grassy margins of oxbow lakes, where it is a widespread and quite common bird. Quite distinctive due to its rather dapper combination of rich brown upperparts and bright creamy-buff underparts, and a long black tail with flashy white tips. This pattern is set off by a black head and gleaming yellow eye. Often gives away its presence by its loud ringing calls. Although often inconspicuous when foraging in dense vegetation, it also sits up in prominent places to call, when it also swings its tail about to show off the white tips.

SEDGE WREN *Cistothorus platensis* 10cm

Also known as the Grass Wren, this is a widespread species throughout the Americas. Found in rough grasslands and marshes in paramo and puna up to about 4000m, where it is usually quite common. Easily identified, as it is the only wren in this habitat with streaked upperparts. The base colour of the upperparts is pale buffy-brown with a paler buff on the underparts. The wings and tail are finely banded. As with so many small birds in this habitat, it usually reveals its presence with its loud, sweet song, which it often delivers from a prominent perch. Has a nasal mewing call.

LONG-TAILED MOCKINGBIRD *Mimus longicaudatus* 29cm

One of the most common and conspicuous birds of the western arid and semi-arid mountain slopes, lowlands and coast (occasionally recorded to nearly 3000m). The only member of the *Mimidae* in Peru, it is a rather tame and very familiar bird. It is instantly recognizable by its long, white-tipped tail, which is often swung about or held aloft in a vertical position. Runs and hops on the ground on its long, strong legs but does not usually fly very far. Often active throughout the day, an indication of its adaptation to the sometimes rather hot and windy environment it inhabits. As its scientific name suggests, it is an accomplished songster and mimic.

SWAINSON'S THRUSH *Catharus ustulatus* 18cm

This little thrush is not a native of Peru but arrives as a migrant from North America in October and November. It often appears in the lowlands, close to the base of the Andean foothills, when bad weather forces the birds to take shelter. Sometimes seen flying over the canopy or crossing rivers in loose flocks, which may cause observers some confusion unless they are familiar with the species elsewhere. Their arrival in the Manu National Park often coincides with that of the Eastern Kingbird *Tyrannus tyrannus*. Very like the scarcer migrant Grey-cheeked Thrush *Catharus minimus*, but has warmer, buffer tones and a very distinct eye-ring.

PLUMBEOUS-BACKED THRUSH *Turdus reevei* 23cm

One of the more strikingly plumaged and instantly recognizable of the Peruvian thrushes. A very distinctive bird with its conspicuous bluish-white eyes and yellow bill contrasting with the lovely blue-grey upperparts and pale grey body plumage with fawn-coloured flanks. The male and female are similar in appearance. A characteristic bird of the deciduous woodlands of the north to about 1500m, and quite common in the forests of the Tumbes region. Typically feeds on the ground or at fruiting trees. Has a sharp call *'seeeeuu'*, and a typical lilting *Turdus* song.

BLACK-BILLED THRUSH *Turdus ignobilis* 22.5cm

This is the common thrush of the eastern lowlands, but is not strictly a bird of the forest. It is easily seen and very familiar to local people, as it is found in habitats such as forest edge, clearings, scrubby areas, orchards and secondary growth. It is also a common garden bird, often seen at feeders, and can become very tame where it is fed and not threatened by domestic cats. Although a rather nondescript bird, it has a lovely, typically thrush-like song. It bears some resemblance to other lowland forest thrushes but is a duller and greyer-brown colour on the upperparts than close relatives such as Hauxwell's Thrush *Turdus hauxwelli*. This is the only lowland thrush that will regularly come close to human habitations.

WHITE-NECKED THRUSH *Turdus albicollis* 22cm

This rather slender and delicate thrush is fairly common in the forests of the Amazonian lowlands and lower Andean foothills but, as with so many forest birds, is more often heard than seen. Its song somewhat resembles that of Hauxwell's Thrush *Turdus hauxwelli*, with which it shares much of its range, but is delivered at a rather slower pace. It is solidly plain brown above with a yellowish bill and eye-ring. The underparts are a rich, plain brown. The throat is heavily streaked with black, bordered below by a conspicuous white crescent. It is a rather shy bird and will very often flee upon approach.

MASKED YELLOWTHROAT *Geothlypis aequinoctialis* 13cm

This is member of the family Parulidae, or New World Warblers, and occurs in marshy habitats with plenty of vegetation on both sides of the Andes. It is widespread in the eastern lowlands and on the Pacific coast, and can be found in pockets of suitable habitat south to the irrigated river valleys around Ica. Typically spends much of its time skulking in reeds, sedges and long grass. Can be seen in coastal

marshes such as the reserve of Pantanos de Villa, on the southern outskirts of Lima. The regional populations in Peru may comprise more than one species, as the males differ in the amount of black on the face, birds east of the Andes having a striking mask. This is much reduced on birds on the Pacific coast. Females lack this feature and have a yellow supraloral stripe.

BANANAQUIT *Coereba flaviola* 11cm

A very common little member of the tanager family. This nectivorous species is one of the birds most often seen raiding hummingbird feeders, often to the point of excluding the hummingbirds themselves. Normally feeds at fruiting trees but commonly enters and sometimes nests inside homes, and will even eat marmalade from the toast on the breakfast plate when one's back is turned! Because of the cheeky nature of this species it is very familiar and its presence is enjoyed by many ordinary people. It is characterized by its striking plumage and short, decurved bill. There are several subspecies, some with a longer bill than others.

WHITE-CAPPED TANAGER *Sericossypha albocristata* 24cm

This large and rather strange and enigmatic tanager of the northern Peruvian highlands is uncommon and infrequently seen. Most often encountered when an observer comes across a noisy flock wandering the forested or even partially-forested slopes of the mountains in search of fruiting trees. They cover huge areas in these expeditions and are reminiscent of a band of jays, usually heard long before being seen. Their loud and far-carrying calls, a shrieking 'cheeeyap', are indicative. Black with a deep purple throat patch, which is larger in the male, its most obvious feature is its startling white cap.

(HIGHLAND) HEPATIC TANAGER *Piranga (flava) lutea* 18cm

This is one of the most common members of the large and diverse tanager family. Formerly simply regarded as just one species, the form *lutea* in Peru is now generally regarded to be a species distinct from the form *flava*. Despite its English name, indicating that it is a

montane species, it does occur in the coastal lowlands south to the Lima area. On the east Andean slope it is generally most likely to be seen between about 1000m and 2500m elevation. The male's red plumage is distinctive, except for possible confusion with the Summer Tanager *Piranga rubra*, an uncommon visitor in the northern winter months. That species has a darker and slightly longer bill and is a paler shade of red. The female is much like the female Summer Tanager, but note the pale bill.

SILVER-BEAKED TANAGER *Ramphocelus carbo* 18cm

One of the most frequently encountered birds of urban and sub-urban areas and secondary growth habitats in the eastern lowlands, where it occurs in almost all gardens. The male is instantly recognizable by the silvery-white sides to its bill and the deep and rather beautiful maroon-red plumage, which often appears black when seen in poor light. The female also has the whitish sides to the bill, although less obvious than in the male. She is a rather uniform rusty-brown colour. Replaced in the Huallaga valley in San Martin and Huanuco by the rather similar but brighter red Huallaga Tanager *Ramphocelus melanogaster*.

PALM TANAGER *Thraupis palmarum* 18cm

One of the commonest tanagers in the lowlands and lower slopes east of the Andes, and also one of the most nondescript. This species is often seen around human habitations but occurs in a variety of habitats where there are trees of any type. As its English name suggests, it has a preference for palms, but this is by no means a strict requirement. Admittedly, however, it is most often seen as a rather nondescript bird flying into a palm tree! Its loud, high-pitched calls are a constant feature of parks and gardens. Its plumage is characterized by its generally rather plain appearance, mainly a dull olive-grey with blackish flight feathers, which give the bird a rather clear-cut bicoloured wing pattern.

BLUE-GREY TANAGER *Thraupis episcopus* 16cm

This is one of the most regularly encountered and conspicuous of Peruvian birds and a common species in and around towns and villages. As with the Palm Tanager *Thraupis palmarum*, it is not found in primary forest but nearly always in secondary growth or cultivated areas. On the west side of the Andes it naturally occurs as far south as the Piura area and up to 1500m, but has been introduced to Lima further south. Easily recognized by its rather lovely blue and grey plumage. The subspecies from the coast, which is illustrated, is rather plain, but the subspecies from the eastern lowlands displays prominent white shoulders.

GOLDEN RUMPED EUPHONIA *Euphonia cynocephali* 11.5cm

The euphonias are members of the tanager family and so-called due to their rather plaintive but melodic calls. All are inhabitants of the forests of the lowlands and foothills. This species occurs up to elevations of over 2500m, and of the ten species of euphonia in Peru this is the only one with a bright blue hood. In the male this contrasts with a black throat and is therefore easily recognized. Like most other euphonia species, the female is a more nondescript bird, being greenish above and yellow below. However, she shares the blue cap of the male and also has a rather unique reddish forecrown, features which make her easily identifiable.

TURQUOISE TANAGER *Tangara mexicana* 14cm

This very easily recognized member of the *Tangara* genus is a bird found in the canopy, edges and clearings of the forests of the humid eastern lowlands, where it is quite common. Its lovely, intricate plumage pattern makes it one of the most distinctive members of the family. The English name is rather odd, as the bird does not have turquoise in its plumage. Instead, it has blackish upperparts and a deep blue throat and flanks, with blackish feathers mottling this colour. The mid-breast and belly are bright yellow. Forages in small parties and does not usually join mixed bird flocks, preferring to remain in the company of its own species.

YELLOW-BELLIED DACNIS *Dacnis flaviventer* 12.5cm

The male of this species is unmistakable, as no other small forest bird remotely resembles it. It has a striking pattern of black on the mantle, wings and tail, and yellow shoulders and rump. It is yellow below with a neat black throat patch. The crown is a rather odd and unique rich olive-green. The head is masked black and highlighted by deep red eyes. The female, however, is a much more obscure bird, being olive-brown above and paler below with brownish mottling and, apart from having red eyes, closely resembles the female Black-faced Dacnis *Dacnis lineata*. The female of that species is characterized by having yellow eyes. This bird is found in the humid eastern lowlands below 500m and is particularly fond of varzea forest.

BLUE DACNIS *Dacnis cayana* 12.5cm

This rather lovely little member of the tanager family is quite common in the lowland forests of the east, occurring up to about 1000m elevation. It is commonly seen in canopy bird flocks or foraging in fruiting trees, but it is also quite a frequent visitor to bird feeders, when available. Apart from both having red eyes, the sexes are very different in appearance, the male being very distinctive with deep, vivid, turquoise-blue on the head and body, and with similarly coloured wing edgings. A black throat patch, black mantle, wings and tail complete the picture. The female (illustrated) is mostly green with a blue head.

RED-LEGGED HONEYCREEPER *Cyanerpes cyaneus* 11.5cm

A very pretty member of the honeycreeper family, characterized by its bright red legs. It also has a beautiful colour pattern of deep, vivid blue underparts, scapulars and rump, black mantle, wings and tail, and a paler and brighter blue crown. It has the typical slender decurved bill of the honeycreepers. The female is greenish-olive above and dull yellowish below. As with the other members of this family, it will readily come to hummingbird feeders and to fruit on bird tables. Although it occurs in the same range and habitat as its close relative, the Purple Honeycreeper *Cyanerpes caeruleus*, it is far less common in Peru than that species.

GREEN HONEYCREEPER *Chlorophanes spiza* 14cm

The largest of the Peruvian honeycreepers and the only one in which the male is bright, shining turquoise-green rather than a shade of blue. The female is very different, a rather uniform green. Commonly visits hummingbird feeders when not at its natural food sources of flowering or fruiting trees. The slightly decurved bill shows a lot of yellow on the lower mandible and base of the upper one. The eye is red. A bird of the lowland and lower foothill forest east of the Andes and the forests of Tumbes in the northwest, up to about 1500m.

PURPLE HONEYCREEPER *Cyanerpes caeruleus* 11cm

The male of this typically beautiful honeycreeper is unmistakable, having deep but bright bluish-purple body plumage and a contrasting black mask, wings and tail. There is no other bird displaying this beautiful colour. However, its most obvious feature is perhaps its extraordinarily thick, bright yellow legs. The very different female is a greenish colour above and streaked below, with an orange-buff throat, blue malar stripe and duller greenish-yellow legs. Both sexes have a slender decurved bill. It is a common member of roving tanager flocks in the eastern lowlands and lower Andean foothills. Most often seen at fruiting trees.

RED-CAPPED CARDINAL *Paroaria gularis* 16.5cm

A common bird along the eastern lowland rivers, streams and lakes, often found close to human habitation. Usually forages low in vegetation close to the water's edge or on floating vegetation, so it is most often seen from boats. It does not skulk, however, and can be rather tame. Nearly always seen in pairs, small groups or family parties. The adult is unmistakable and rather pretty, with its bright red head, black throat patch, glossy blue-black upperparts and gleaming white underside. The juveniles are rather less obvious, being browner above and with a buffish-brown head and paler buffish throat and upper breast.

YELLOW-HOODED BLACKBIRD *Agelaius icterocephalus* 18cm

Another typical black and yellow member of the *Icteridae* family from further north, but one which only just makes it into Peru. It occurs only in the Amazonian lowlands in the northeastern province of Loreto. Its preferred habitat is densely vegetated marshes and areas of cane, and even here it is generally rather uncommon. It will usually forage at low level but sometimes perches on top of vegetation, when it can be very conspicuous. The male is quite unmistakable, its bright yellow head contrasting with the black body. The female is very different, being a rather dull blackish-olive with paler edges and darker centres on the mantle. It has a dull, yellowish breast and supercilium.

YELLOW-TAILED ORIOLE *Icterus mesomelas* 21.5cm

The only yellow oriole in Peru, with conspicuous bright yellow outer feathers to its black tail, a vivid yellow head, underparts, shoulders and rump. The contrasting black throat and upper breast, mantle, and wings make this one of the most conspicuously plumaged Peruvian birds. It occurs in woodland and scrub in the northern coastal lowlands, south as far as Lambayeque and also in the upper Marañon valley, but is not particularly common. Occurs alongside the White-edged Oriole *Icterus graceannae*, but that species sports a diagnostic white patch on the tertials and has an all black tail.

GIANT COWBIRD *Scaphidura oryzivora* 12-15cm

Another highly iridescent member of the *Icteridae* family. Very often seen from boats travelling along the waterways of the eastern lowlands, and also occurs along the coast in the Tumbes region. The males display by strutting on their long legs with inflated breasts, which accentuate their large size. In flight this species has a powerful loping action which, once recognized and learnt, is very distinctive. The males can also be easily identified by being much larger than the females. As is usual with the cowbirds, this species is a brood parasite and, as the largest of them all, parasitises oropendolas and caciques. The females can often be seen hanging around the vicinity of a colony waiting for a chance to deposit an egg.

YELLOW-RUMPED CACIQUE *Cacicus cela* 27cm

A common bird of the eastern lowland humid zone, but also occurs in the Tumbes region in the far north, where it is less common. Very often seen breeding alongside oropendolas, and has rather similar hanging nests. This species is also often closely associated with human habitation, where it nests close to houses, especially along watercourses, although not exclusively. It has a striking black and yellow plumage pattern, and a distinctive, loose flight action, looking sleek and long-tailed. This combination makes this an unmistakable bird in its environment. Males have a loud and complex song, which is delivered as the bird bends over and ruffles the splayed feathers of its back to attract attention to its bright yellow rump.

CRESTED OROPENDOLA *Psarocolius decumanus* 45cm

One of six species in Peru, this is only black-bodied oropendola. A bird of the forested Amazonian lowlands, it has a long and pointed, ivory-white bill. Its 'crest' is barely visible, being just a single filament. A very striking bird, with brilliant yellow outer tail feathers, as with other members of the group. Also in common with other oropendolas, it is named after its habit of building pendulous, hanging nests in conspicuous colonies. Very often these are in isolated trees in clearings or on riverbanks, well away from the forest edge and where they are afforded protection from predation by monkeys. Often seen flying across rivers in loose flocks en route to a food source or returning to the colony after a foray.

GOLDEN-BILLED SALTATOR *Saltator auratiirostris* 20cm

A common and conspicuous bird of the high Andes up to an altitude of about 3000m. Occurs in various dry and scrubby habitats and is regularly seen in and around Cuzco in the southern highlands. As its English name suggests, this species is characterized by its large and brightly coloured bill, which is orange-yellow in the male and somewhat duller and less orange in the female. Both sexes also have a very beautiful plumage pattern with a striking black mask, which continues down and around the breast. This contrasts with the white stripe on the head and the white throat, which varies in size depending where one sees the bird – it is smaller and less obvious in the north of the species's range.

BUFF-BRIDLED INCA-FINCH *Incaspiza laeta* 14.5cm

This handsome inca-finch, like all the other four species, is not only endemic to Peru but restricted to the northern part of the country, where it is quite common in its required habitat of Bombax woodland and xerophytic scrub. Most likely to be seen in the Marañon valley, between about 1500m and 2750m, and can be seen along the road from Celendin to Leimabamba on either side of the valley above Balsas. It is a rather flighty species, so can be difficult to approach. It has a typical inca-finch pattern of rufous and grey with a black throat, but is the only member of this family to have a conspicuous buff malar stripe.

GREY-WINGED INCA-FINCH *Incaspiza ortizi* 16.5cm

Another of the endemic birds of the north of the country, where it lives in dense scrub in rather arid areas, as do its other close relatives. This very localized species can be seen on the west side of the Marañon valley at about 2000m elevation, in the area around Hacienda Limon on the road from Celendin to Balsas, but it is rather uncommon and elusive. Only known from a handful of other sites in Piura and Cajamarca, one of which is La Esperanza, about 5km northeast of Santa Cruz. Also found within the first few kilometres along the road from Huancabamba to Sapalache. It is the only inca-finch that does not have rufous on the upperparts.

WHITE-WINGED DIUCA-FINCH *Diuca speculifera* 19cm

This large finch is a bird of high altitudes and a characteristic bird of the puna zone, where it occurs from about 4500m to 5350m, right up to the snow line. Usually seen feeding in short grassy or boggy areas, or in the highest, sparsely vegetated habitats. A very distinctive bird, being mainly a pale, plain, soft grey colour. There is an obvious white crescent under the eye and a clear-cut white throat. The belly is white, but perhaps the white patches at the base of the flight feathers are the most characteristic feature. These contrast with the somewhat blacker wings and are the bird's most obvious feature in flight.

HOODED SISKIN *Carduelis magellanicus* 12.5cm

One of the commonest members of the finch family in Peru, especially in agricultural areas of the highlands. Also occurs commonly along the Pacific coast. There are several different races in Peru, but males all have the typical black hood. Females are less handsome, lacking both the male's black hood and the bright yellow band in the wing. However, there are several other very similar species, which may cause some confusion, particularly the Olivaceous Siskin *Carduelis olivacea*. The form *paula* from the northwest occurs sympatrically with the rare Saffron Siskin *Carduelis siemiradzkii* but that species has a brighter, plain golden-olive mantle.

LINED SEEDEATER *Sporophila lineola* 11.5cm

A common migrant to the eastern lowlands during the austral winter. As is typical of the seedeater family, the male is rather different from the female, being strikingly patterned, whereas the female is an anonymous plain buffy-brown above and a paler buffy shade below. The male is black above and white below with a black throat, broad white malar stripe and a white median crown-stripe. It is very similar to Lesson's Seedeater *Sporophila bouvronides*, but that species lacks the crown stripe and visits during the boreal winter. Typically occurs in small flocks, in which there is usually only one male in a group of perhaps seven or eight females.

SAFFRON FINCH *Sicalis flaveola* 14cm

Perhaps the most familiar of the nine species of *Sicalis* finches in Peru, this bird occurs commonly in agricultural areas and scrub along the arid coastal plain south to Ancash. Also extends up the western Andean slope to a height of about 1000m. It is an easily recognized and conspicuous bird in many lowland areas, and around towns and villages, including gardens. This bright yellow finch, with its orange forehead, is arguably the easiest of the *Sicalis* finches to identify, being the only one with such plumage within its range. The female is similar but a little duller. The young birds are grey-brown above with dark streaking and pale greyish below, but are characterized by a yellow band across the breast.

GRASSLAND YELLOW-FINCH *Sicalis luteola* 12.5cm

Usually quite a common species where it occurs. As its English name indicates, it is typically found in grassy areas such as pastureland, both along the coast and in the highlands to about 3000m. Particularly likes marshy places and the proximity of water, shunning arid areas. Can be erratic in its occurrence, depending on climatic conditions affecting its habitat. Generally yellow below and brownish above, streaked darker. The head is marked by a yellow supercilium, lores, and a patch just below the eye. Has a narrow, darker malar streak, which produces a rather vague yellow moustachial stripe. The female is generally duller and paler below.

RUSTY-BELLIED BRUSH-FINCH *Atlapetes nationi* 17cm

This brush-finch is a member of a genus which is amply represented in Peru with sixteen species in total, of which six are endemic to the country. Fairly common in dry, bushy habitats and Polylepis woodland on the arid western slopes of the Andes, from Lima south to Arequipa, between about 2000m and 4000m elevation. Generally grey above with a darker and browner crown and blackish 'face' and chin. The lower throat and malar area is white. It has a rather paler grey breast, which gently blends into a pale, rufous belly, becoming richer on the vent.

YELLOW-BROWED SPARROW *Ammodramus aurifrons* 13cm

In many ways this bird replaces the Rufous-collared Sparrow *Zonotrichia capensis* in marginal habitats and agricultural areas, such as pasture, at lower elevations east of the Andes. Likely to be present anywhere with grassy areas in the Amazonian lowlands and up to about 1000m. Not only is it a common bird here but its buzzy, trilling song is very frequently the birdsong most often heard from riverbanks. It sounds rather insect-like and can be heard even in the heat of the day, when most other birds have fallen silent. A rather nondescript, small brown streaky bird, but always has some degree of yellow on the head, usually on the lores, eye-ring and short supercilium.

RUFOUS-COLLARED SPARROW *Zonotrichia capensis* 14cm

Perhaps the most ubiquitous small bird in Peru, being found in all highland regions above about 1500m on the east Andean slope, and all the way down to sea level in the west. Replaced by the Yellow-browed Sparrow *Ammodramus aurifrons* in the humid eastern lowlands. Occurs in urban and rural areas alike, where its pretty, high-pitched and whistled song is heard almost constantly from dawn till dusk. Very common in Lima at sea level and in Cuzco at over 3000m. The adult is very distinctive, with its rusty hind-neck, grey head with black eyestripes and sides of the peaked crown. The brown, streaky juveniles lack these features but usually show some trace of the rusty collar.

GLOSSARY

Clay Lick Also known as a collpa. A place where parrots congregate to consume minerals, which assist their digestion. Such sites are usually on riversides where the banks are eroded and exposed. Also used by parrots as a meeting point for social interaction. Rather like 'going down the pub' for a parrot!

Congener A close relative of another particular species, and specifically a member of the same genus.

Ear coverts The 'cheeks' of a bird.

Endemic A species restricted to a defined area, usually a particular country. Peru has more than 100 national endemics, ie. bird species that are found nowhere else in the world.

Fluvial Bogs These occur in high Andean areas in the 'puna zone', where meltwater runs off and spreads into flat areas, rather than following a stream bed. This is the specific habitat for the beautiful Diademed Sandpiper-Plover.

Genus A grouping of closely related species.

Lek A specific area (also a verb, to lek) used by some species as a place in which to display. Often a noisy and rather spectacular event, particularly where cotingas such as the Cock-of-the-rock are lekking. Some hummingbirds also lek.

Malar The area just below the ear coverts.

Mantle The 'back' of a bird.

Nectivorous Feeding largely on sugar, including ripe fruit, flowers or other sugar sources (such as man-made solutions at feeders).

Polylepis An evergreen tree with reddish, flaky, paper-like bark and which grows above the normal tree line on high and often steep slopes. It forms a unique high-altitude woodland habitat for many bird species. Drastically threatened by cutting in many areas.

Puna The zone of grassy habitats above the tree line in the high Andes and the Altiplano. Interestingly, puna is also the Quechua name for altitude sickness.

Supercilium The stripe that is often a distinct feature and which runs over the eye and above the eyestripe. Usually pale and obvious.

Tertials The innermost large wing feathers that cover where the wing joins a bird's body.

Traplining The feeding strategy utilized by some species of hummingbird, by which they exploit widely dispersed, nectar-rich flowers, visiting them on a regular circuit. By the time a flower is visited on a subsequent round it has replenished its nectar supply.

Xerophytic scrub A habitat common in the drier areas of the Peruvian highlands, especially in the north. Typified by stunted trees, thorny, often leafless shrubs and cacti in particular.

FURTHER READING

Clements, James & Shany, Noam. *A Field Guide to the Birds of Peru* (Ibis, 2001).

Forsyth, A. & Miyata, K. *Tropical Nature: Life and Death in the Rain Forests of Central and South America* (Simon & Shuster, 1995).

Harrison, Peter. *Seabirds: An identification Guide* (Christopher Helm, 1991).

Hilty, S. *Birds of Tropical America* (University of Texas Press, 2005). Highly recommended as a book that anyone will enjoy, regardless of level of experience. Read it on the plane!

Hilty, S. & Brown, W.L. *A Guide to the Birds of Colombia* (Princeton, 1986).

Jaramillo, A. *Birds of Chile* (Princeton Field Guides, paperback, 2003).

Kricher, John. *Neotropical Companion* (Princeton University Press, 1999).

Ridgely, Robert S. & Tudor, Guy. *Birds of South America* (two volumes) (Oxford University Press, 1989, 1994).

Ridgely, Robert S. & Greenfield, P.J. *The Birds of Ecuador* (two volumes) (Christopher Helm, 2001).

Valqui, Thomas. *Where to Watch Birds in Peru* (Valqui, 2004; www.granperu.com).

Walker, Barry. *Field Guide to the Birds of Machu Picchu* (2002).

INDEX